SAE International's Dictionary of Vehicle Accident Reconstruction and Automotive Safety

In Remembrance of Linda Ayres-DeMasi

In the quiet solitude of the written word, you found your place of grace. Your keen eye, your unwavering commitment to clarity, and your unending pursuit of excellence breathed life into the pages of countless manuscripts.

Your gentle guidance and your tireless encouragement propelled authors forward when doubt and uncertainty loomed large. Without you, many books would have remained unpublished.

You may have left this world but your spirit lives on in the books that bear your mark, in the hearts of the authors whose journeys you guided and in the memory of everyone inspired by your professionalism.

With heartfelt gratitude,

Sherry Nigam
Publisher, SAE Books
SAE International

SAE International's Dictionary of Vehicle Accident Reconstruction and Automotive Safety

BY
R. MATTHEW BRACH, PhD PE

SAE INTERNATIONAL®

Warrendale, Pennsylvania, USA

400 Commonwealth Drive
Warrendale, PA 15096-0001 USA
E-mail: CustomerService@sae.org
Phone: 877-606-7323 (inside USA and Canada)
724-776-4970 (outside USA)
FAX: 724-776-0790

Library of Congress Catalog Number 2023946934
http://dx.doi.org/10.4271/9781468605945

ISBN-Print 978-1-4686-0593-8
ISBN-PDF 978-1-4686-0594-5
ISBN-epub 978-1-4686-0595-2

To purchase bulk quantities, please contact: SAE Customer Service

E-mail: CustomerService@sae.org
Phone: 877-606-7323 (inside USA and Canada)
724-776-4970 (outside USA)
Fax: 724-776-0790

Visit the SAE International Bookstore at books.sae.org

Publisher
Sherry Dickinson Nigam

Development Editor
Publishers Solutions, LLC
Albany, NY

Director of Content Management
Kelli Zilko

Production and Manufacturing Associate
Michelle Silberman

Contents

Preface

The starting point for this document is the glossary contained in *Vehicle Accident Analysis and Reconstruction Methods* by Brach, Brach, and Mason [1]. The initial version of that glossary, which appeared in the first edition published in 2005, borrowed from the efforts of the now disbanded SAE Accident Investigation and Reconstruction Practices Committee (AIRP). That group collected and assembled a glossary of terms relevant to the field of accident reconstruction but never published a recommended practice.

To satisfy the scope and intent of breadth of the present document, that glossary was expanded to include more terms from the field of accident reconstruction and many terms from the field of automotive safety. A significant number of the terms from automotive safety came from the field of biomechanics with the interpretation of automotive safety as it relates primarily to the occupants of a vehicle. In contrast, the terms related to accident reconstruction pertain primarily to the vehicle.

Unlike most glossaries, this document includes references. The intent of providing references is not only to recognize the work of the authors of the sources, but also to give the user of this document both a relevant definition and a source of additional information should the reader require more than is provided here. Many of the terms came from the vast number of SAE standards and recommended practices dealing with accident reconstruction, automotive safety, and related automotive topics. Other terms came from the glossaries of various books. Where used, these references are identified.

Apologies are provided to users who do not find their desired term and its definition. The task of assembling this document was daunting and inherently never complete as the automotive industry continues to evolve. Users are invited to send the author recommendations for changes and suggestions for terms to include in the future versions of this document at the following email address: customerservice@sae.org.

Acknowledgments

The author would like to thank Linda DeMasi and Bruce Sherwin of Publisher's Solutions, LLC, and Sherry Nigam of SAE International for their help in assembling this publication. The author is indebted to Mr. Richard Mink for reading the entire manuscript and making numerous useful suggestions and modifications that improved the final product.

A

A, B, C, D Pillars and Posts

The vertical pillars and posts of a vehicle that form the major vertical structural members of the body, which typically support the vehicle's roof (see Figure 1). Pillars typically are at window height; posts are below window height. From front to rear, the A post/pillar is the most forward member, the B post/pillar is the second most forward vertical member, and so on.

FIGURE 1 Vehicle pillars and posts.

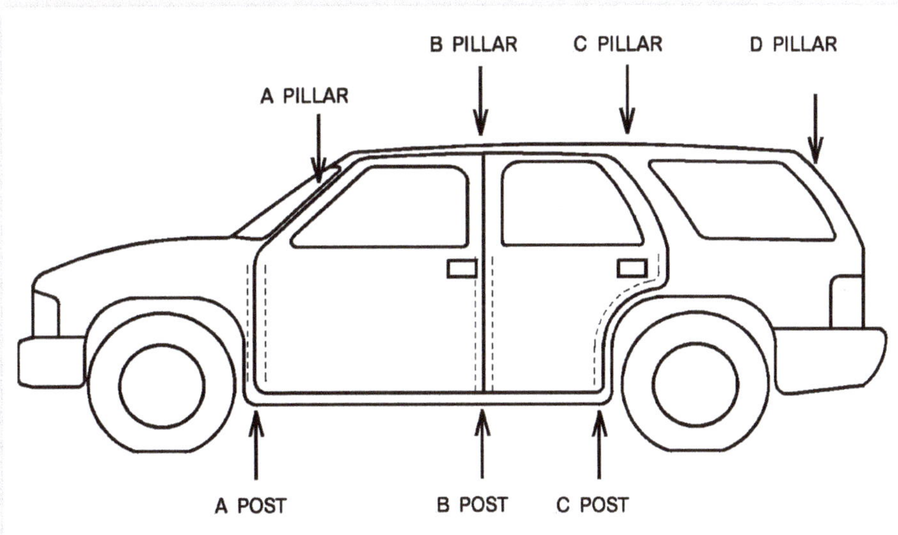

AAAM

Association for the Advancement of Automotive Medicine. The AAAM is a multidisciplinary organization dedicated to motor vehicle crash injury prevention and control. This organization was involved in the creation of the Abbreviated Injury Scale (AIS). See Abbreviated Injury Scale.

AACN

Automatic Advanced Crash Notification. General Motors' term for an automatic collision notification system (ACN). See ACN.

AADT

Average annual daily traffic. The total volume of vehicle traffic on a highway or roadway for a year divided by 365 days. AADT is a useful and simple measurement of how busy a road is.

AASHTO

American Association of State Highway and Transportation Officials.

ABA

American Bus Association. The ABA supports the motorcoach, group tour, and travel association industry. Also stands for active brake assist.

Abbreviated Injury Scale (AIS)

A numerical rating system for quantifying the severity of injuries to an accident victim. The rating scale is as follows:

CODE	CATEGORY
1	Minor
2	Moderate
3	Serious
4	Severe
5	Critical (survival uncertain)
6	Maximum (virtually unsurvivable)
9	Unknown

For more information, see **J885_FEB2011** [4].

Abduction

Motion of a human body appendage away from the body's midline. For more information, see *Basic Biomechanics of the Musculoskeletal System* [3].

ABS

See Antilock Brake System.

ABS Configuration

The configuration name is based on the number of modulator valves and sensors used on a vehicle equipped with an antilock brake system. A common configuration used on commercial trucks and tractors is 4S/4M (four sensors and four modulator valves) For more information, see *Commercial Vehicle Braking Systems: Air Brakes, ABS and Beyond* [5].

ABS Modulator Valve

For vehicles that have an air-brake system and are equipped with an antilock brakes system, an electropneumatic control valve that contains the solenoids used for controlling air pressure. It can precisely modulate air pressure during an ABS-controlled brake application using signals received from the ABS ECU. For more information, see *Commercial Vehicle Braking Systems: Air Brakes, ABS and Beyond* [5]. The modulation of the air pressure is generally handled individually chamber by chamber to limit wheel slip at a given wheel.

ACAT

Advanced collision avoidance technology.

ACC

See Adaptive Cruise Control.

Acceleration

Change in linear or angular velocity with respect to time. Acceleration is a vector quantity possessing both magnitude and direction. Deceleration is acceleration with negative magnitude.

Accelerator Pedal Position

The angular displacement of the accelerator pedal relative to its position at idle. For more information, see **J670_JUN2022** [6]. An event data recorder (EDR) generally reports the pedal position as a percentage, where 100% represents a fully depressed accelerator pedal.

Accelerometer

An electromechanical sensing device with an output signal in volts proportional to the accelerations acting on the sensor.

Accident Investigation (AI)

The process of observation, acquisition, and documentation of physical evidence and other information regarding an accident, incident, or crash.

Accident Reconstruction (AR)

A procedure carried out with the specific purpose of estimating in both a qualitative and quantitative manner how an accident occurred, using engineering, scientific, and mathematical principles. AR is based on physical evidence obtained through an accident investigation.

Accident Scene

A place where a traffic accident occurs, both during and immediately following the accident, and before vehicles and participants have departed. See Accident Site.

Accident Site

A place where a traffic accident occurred, after vehicles and participants have departed the scene. See Accident Scene.

Accident (Vehicle)

An event in which one or more vehicles undergo unexpected action(s), usually involving contact with another vehicle or other object, producing injury, death, and/or property damage; an accident is an unstable situation that includes at least one harmful event. See Crash.

Ackerman Steer Angle

The angle whose tangent is the wheelbase (WB) divided by the radius of turn (see Figure 2). For more information, see *Road Vehicle Dynamics* [8].

FIGURE 2 Basic Ackerman steering diagram [7].

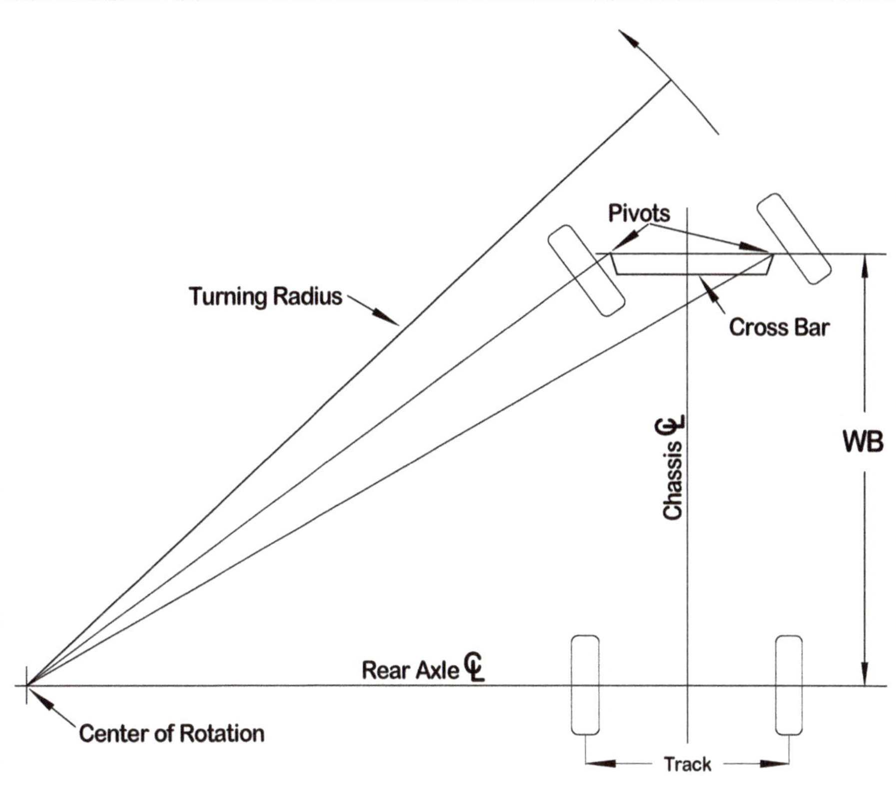

Ackerman Steer Angle Gradient

The rate of change of the Ackerman steer angle with respect to the change in steady-state lateral acceleration on a level road at a given trim and test conditions. For more information, see *Road Vehicle Dynamics* [8].

Ackerman Steering System

The degree at which the steering arms are set varies with the wheelbase of the unit and compensates for the difference in turning radius between the two front wheels. By closely complying to these angles, a minimum of tire wear will be experienced when executing the turns. For more information, see *Motor Truck Engineering Handbook* [7]. The design of a vehicle's steering linkages such that the steer tire on the outside of the turn articulates less than the steer tire on the inside of the turn for each amount of steering input, thereby allowing the outside tire to follow the larger-radius path along the outside of a turn.

ACM

Airbag control module. An electronic control module for airbags and related restraint systems. See Event Data Recorder; RCM; and SDM.

ACN

Automatic collision notification system. A system that can help emergency responders to better and more quickly assess whether a motorist involved in a vehicle accident needs emergency care. An example is General Motors' OnStar system.

Acoustic Levels

See Sound Level.

Active Fault

A current failure of a vehicle system in need of repair. Examples are failure of a sensor circuit, a broken wire, or an inoperative ABS valve solenoid. For more information, see *Commercial Vehicle Braking Systems: Air Brakes, ABS and Beyond* [5].

Active Safety System

Active safety systems sense and monitor conditions inside and outside a vehicle to identify perceived present and potential dangers to the vehicle, occupants, and/or other road users, and automatically intervene to help avoid or mitigate potential collisions via various methods, including alerts to the driver, vehicle system adjustments, and/or active control of the vehicle subsystems (brakes, throttle, suspension, etc.). For more information, see **J3016_JUN2018** [9].

ADA

Americans with Disabilities Act. The ADA, passed by the United States Congress in 1990, is a civil rights law that prohibits discrimination based on disability.

Adaptive Cruise Control (ACC)

An enhancement of conventional cruise control (CCC) systems that allows the ACC-equipped vehicle to follow a forward vehicle at a preselected time (or distance) gap, up to a driver selected speed, by controlling the engine, powertrain, and/or service brakes. For more information, see **J2399_OCT2021** [10].

ADAS
Advanced driver assistance system. Any automotive system or group of systems of a variety of electronic control systems that assists an automobile operator in driving or parking.

Adduction
Motion of a body appendage toward the body's midline. For more information, see *Basic Biomechanics of the Musculoskeletal System* [3].

ADR
Accident data recorder. See Event Data Recorder.

ADS
See Automated Driving System.

ADS-Dedicated Vehicle
A vehicle designed to be operated exclusively by a level 4 or 5 ADS for all trips within its given operational design domain (ODD) limitations (if any). For more information, see **J3216_JUL2021** [11].

AE
See Algorithm Enable.

AEB
See Automatic Emergency Braking. Also stands for autonomous emergency braking.

Aerodynamic Angle of Attack
The angle between a vehicle's x-axis and the trace of the resultant air velocity vector on a vertical plane containing the vehicle x-axis. For more information, see *Road Vehicle Dynamics* [8].

Aerodynamic Sideslip Angle
The angle between the traces on a vehicle's x-y plane of the vehicle x-axis and the resultant air velocity vector at some specified point in the vehicle. For more information, see *Road Vehicle Dynamics* [8].

AES

See Automatic Emergency Steering.

AF

Axle-to-frame. For a heavy truck, the distance from the centerline of the rear axle(s) to the end of the chassis frame (see Figure 3).

FIGURE 3 Dimensional nomenclature for a truck tractor.

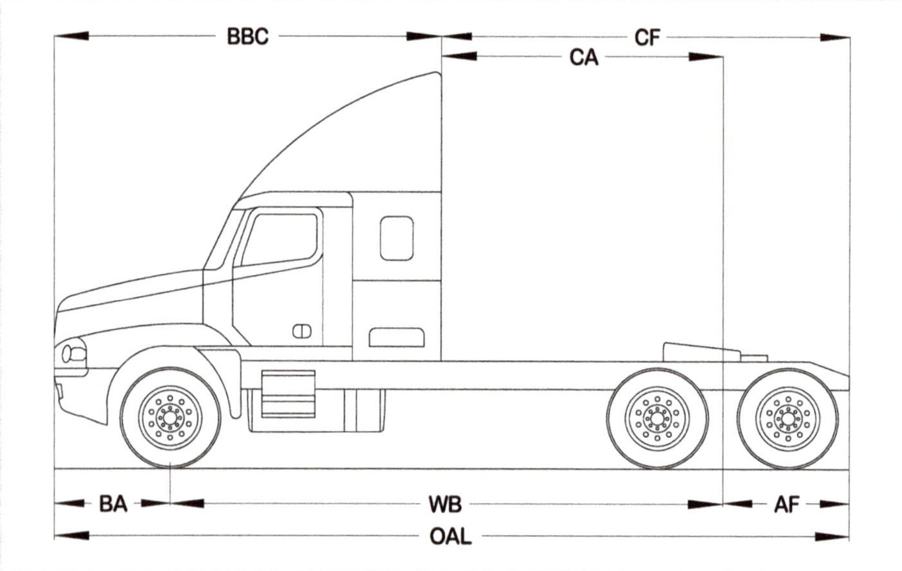

AFLS

Advanced front lighting system. A system on an automobile that can adjust the direction and intensity of the vehicle headlights.

Aftercooler

See Intercooler.

AFV

Alternative fuel vehicle. An AFV is a motor vehicle that runs on an alternative fuel rather than on traditional petroleum-based fuels. The term can also refer to vehicles that have a propulsion system that is not solely petroleum based, such as a hybrid electric vehicle.

Aggressivity

The inertial and structural properties and characteristics of a vehicle that relate to the potential for severe injuries to the occupants of the other vehicle(s) in a crash.

Agonistic Muscles

Muscles that initiate and carry out motion. For more information, see *Basic Biomechanics of the Musculoskeletal System* [3].

Agricultural Commodity Trailer

A trailer designed to transport bulk agricultural commodities from harvest sites to process or storage sites. For more information, see 49 CFR §393.5 Definitions [12].

AHBC

Adaptive high beam control. A system that provides for the precision control of the distribution of light from the headlight system on a vehicle.

AHP

Accelerator heel point. For more information, see **J826_JUN2021** [13] and **J1100_NOV2009** [14]. Also see Figure 26.

AI

See Accident Investigation.

Airbag

A device in the interior of a vehicle that inflates and acts between an occupant and an interior vehicle surface to prevent injury in a crash. See Supplemental Restraint System.

AIS

See Abbreviated Injury Scale.

ALA

Active lane assist.

ALC

Adaptive light control. A vehicle system designed to improve nighttime visibility without adversely affecting the vision of the drivers of other vehicles.

Algorithm Enable (AE)

A programmed trigger for a specific airbag control module (ACM) when the ACM begins the seat belt pretensioner/airbag deployment decision-making algorithm.

Aligning Moment (Aligning Torque, Self-Aligning Moment)

The component of the tire moment vector tending to rotate the tire about the Z_T axis (see Figure 4). For more information, see *Road Vehicle Dynamics* [8] and **J2047_NOV2019** [15].

FIGURE 4 Wheel force and moment nomenclature.

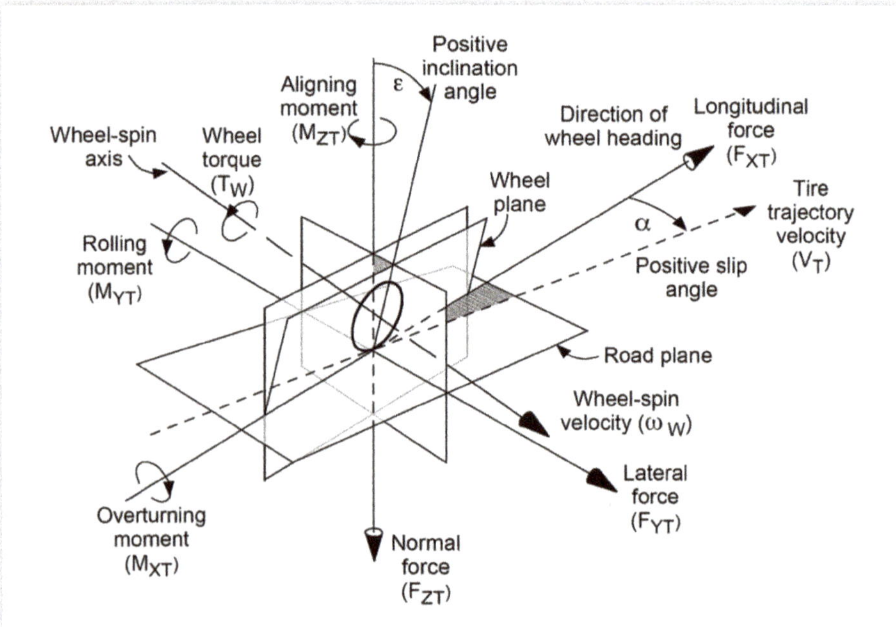

All Belts to Seat (ABTS)

A seat belt assembly in which all the anchorage points are secured to the seat assembly. Also known as Seat integrated restraints (SIR), belt in seat (BIS), or integrated restraints. For more information, see **J1803_JAN2013** [16].

All-Electric Range

The distance an electrified vehicle can travel without using any fuel (gas, diesel, natural gas, etc.) for propulsion or for performing any other functions of the vehicle. Note that the all-electric range must be accompanied with a reference to the drive cycle used to determine the range. For more information, see J1715_SEP2022 [17].

AMA

American Motorcyclist Association. An organization that pursues, protects, and promotes the interests of motorcyclists, and is the premier sanctioning organization for motorcycle racing and other competitive events in the United States. For more information, see **J3133_SEP2019** [18].

Angular Acceleration

The time rate of change of angular velocity.

Angular Momentum

The angular momentum of a body about its axis of rotation is the moment of its linear momentum about that axis. For more information, see *Road Vehicle Dynamics* [8].

Angular Velocity

The time rate of change of rotational displacement.

Animation

The process by which the movement of objects is illustrated.

ANSI

American National Standards Institute. ANSI oversees the development of voluntary consensus standards for products, services, processes, systems, and personnel in the United States.

Antagonistic Muscles

Muscles that oppose the actions of the agonistic muscles. For more information, see *Basic Biomechanics of the Musculoskeletal System* [3].

Anthropometry

An empirical science defining the physical measures of a person's size and form. For more information, see *Occupational Biomechanics* [19].

Anthropomorphic Test Device (ATD)

Test dummies that are used in crashes instead of a living human; an example is the Hybrid III dummy. For more information, see *Motor Vehicle Collision Injuries: Biomechanics, Diagnosis, and Management* [20].

Anti-Compounding Valve

In vehicles with air brakes, a valve that prevents the application of the service and the spring brakes at the same time. For more information, see *Commercial Vehicle Braking Systems: Air Brakes, ABS and Beyond* [5].

Anti-Dive

A design or suspension parameter that reduces the amount of nose-down pitching or front suspension compression that occurs during moderate to heavy breaking. For more information, see **J3133_SEP2019** [18].

Antilock Brake System (ABS)

A closed-loop control system designed to regulate brake torque to limit the tire longitudinal slip ratio during brake application. This is typically accomplished by transducers for sensing wheel rotation and an actuation system under control of a computer that regulates brake pressure. For more information, see **J670_JUN2022** [6].

AOI

See Area of Impact.

Apply Time

With an initial service reservoir pressure of 100 psi, the measurement in time from the movement of the service brake control for the brake chamber to reach 60 psi. For more information, see *Commercial Vehicle Braking Systems: Air Brakes, ABS and Beyond* [5].

Approach Angle

For a motorcycle, the smallest angle, in a planar side view of a motorcycle, formed by the surface on which the motorcycle is standing and a line tangent to the front tire circumference and touching any part of the motorcycle forward of the front tire (such as the front fender). The notion of approach angle has also been applied to heavy trucks and SUVs. For more information, see **J3133_SEP2019** [18].

Approach Speed

Speed of a vehicle just prior to the first significant event such as contact in an accident. See Closing Speed.

APS

Automatic parking system. An APS is a system designed to minimize the space required for parking cars.

APU

See Auxiliary Power Unit.

Aquaplaning

See Hydroplaning.

AR

See Accident Reconstruction.

Area of Impact (AOI)

Area encompassed by the interface between colliding objects projected onto the roadway. See Point of Impact.

Articulated Vehicle

A vehicle comprised of two or more distinct, interconnected bodies (masses) such as a truck tractor and a semitrailer.

Articulation Point

The instant center of rotation of the trailer relative to the towing vehicle. For more information, see **J670_JUN2022** [6].

ASA

See Automatic Slack Adjuster.

Asphalt

See Bituminous Pavement.

ATA

American Trucking Associations. In the United States, a trade organization for the trucking industry.

ATC
Automatic traction control.

ATD
See Anthropomorphic Test Device.

Attachment Face
On a wheel, the surface of the disc that contacts the hub face. This surface is often considered the datum for axial rim measurements. For more information, see **J2047_NOV2019** [15]. See item 7 in Figure 19.

Attachment Hardware (seat belt)
Attachment hardware encompasses any or all hardware designed for securing the webbing of a seat belt assembly to a motor vehicle. If the seat belt is attached to a seat, the seat is not considered attachment hardware. For more information, see **J1803_JAN2013** [16].

Automated Driving System (ADS)
The hardware and software that are collectively capable of performing the entire dynamic driving task (DDT) on a sustained basis, regardless of whether it is limited to a specific operational design domain (ODD); this term is used specifically to describe a level 3, 4, or 5 driving automation system.

Automatic Disconnect
A device in any type of xEV that opens the primary high voltage circuit or circuits of a vehicle, and that is activated by automatic means, such as a crash response or the operation of the vehicle ignition switch. For more information, see **J2990_NOV2020** [21].

Automatic Emergency Braking (AEB)
Combines advanced driver assistance systems and subject vehicle braking to slow a vehicle to avoid or mitigate a collision. For more information, see **J3157_FEB2019** [22].

Automatic Emergency Steering (AES)
Detects potential collisions with a vehicle ahead and automatically steers to avoid or lessen the severity of impact. Some systems also detect pedestrians or other objects. For more information, see **J3063_MAR2021** [23].

Automatic Locking Retractor (ALR)

A retractor incorporating a positive self-locking mechanism that remains engaged, locked, and capable of withstanding restraint forces once the webbing has initially been extracted. This locking mechanism is reset when the webbing is allowed to retract back into the retractor. For more information, see **J1803_JAN2013** [16].

Automatic Slack Adjuster

On a vehicle equipped with pneumatic brakes, a component that provides the mechanical link between the brake chamber and push rod and the camshaft on a drum brake system that also provides, during normal use, for continuous and automatic adjustment of the length of the brake stroke.

Auto-Stabilizing

The dynamic behavior of a motorcycle's front fork assembly, whereby it tends to return to a centered position if proceeding in a straight line or to a trim condition if the motorcycle is in a turn. For more information, see **J3133_SEP2019** [18].

Auxiliary Power Unit (APU)

A device used to generate propulsion power but without a direct mechanical connection to the drive wheels. Some examples of auxiliary power units are internal combustion engines connected solely to an electric generator, an engine connected solely to a hydraulic pump, or fuel cells. For more information, see **J1715_SEP2022** [17].

On heavy trucks, APUs are typically used to provide electrical power and HVAC functions while the vehicle is parked with the primary engine not running.

AV

Automated vehicle. Automated vehicles are those in which at least some aspect of a safety-critical control function (e.g., steering, throttle, or braking) occurs without direct driver input.

Average Acceleration

Calculated by dividing the change in velocity of an object by the time that is required to gain that velocity. For more information, see *Road Vehicle Dynamics* [8].

Average Driver Workload

The amount of physical and mental activity that is required to perform a particular task or set of tasks while driving during the time it takes to complete them. For more information, see **J3077_DEC2015** [24]. See Driver Workload; Instantaneous Driver Workload; Overall Driver Workload.

Avulsion

Traumatic tearing of body tissue. For more information, see *Principles of Biomechanics* [25].

AWD

All-wheel drive.

Axle

(1) An essentially rigid member used to connect wheels on opposite sides of a vehicle at a given longitudinal position. (2) The imaginary line connecting the wheel centers on opposite sides of a vehicle at a given longitudinal position. (3) Less rigorously, the assemblage of the suspension, wheels, tires, brakes, and all other components that move with the wheels at a given longitudinal position of the vehicle. For more information, see **J670_JUN2022** [6].

Axle, Pusher

Nonpowered rear axle located ahead of the drive axle(s). For more information, see *Motor Truck Engineering Handbook* [7].

Axle, Tag

Nonpowered rear axle located behind the rearmost drive axle that lengthens the wheelbase of a vehicle.

B

BA

Bumper-to-axle. For a truck tractor, the distance from the front bumper to the front axle. See Figure 3.

BAC

Blood alcohol concentration. Refers to the concentration, in percent by volume, of alcohol in a person's blood.

Backlite

The rear or back window that spans from the driver's side to the passenger's side of a vehicle (see **Figure 5**).

FIGURE 5 Light vehicle backlite.

VetalStock/Shutterstock.com.

Backlite Header

The structural body member that connects the upper portions of the rearmost driver and passenger pillars. In a light vehicle, the backlight header forms the top edge of the backlite (back window).

Ball Joint (Rod End)

A type of mechanical connector consisting of one component having a spherical end and the other component having a mating hemispherical socket into which the spherical end is inserted and contained. This geometry allows for rotational but not translational motion. Ball joints are generally used in the steering system of a vehicle to connect a link or control arm to the knuckle or used at each end of a tie rod (see Figure 6).

FIGURE 6 Ball joint.

Varavin88/Shutterstock.com.

Barrier Equivalent Velocity (BEV)

The forward speed and corresponding kinetic energy with which a vehicle contacts a flat, fixed, rigid barrier at 90° with no rebound. See also Energy Equivalent Speed (EES).

BAS

See Brake Assist System.

Battery Electric Vehicle (BEV)

A vehicle that receives its on board propulsion power solely from batteries, unlike a hybrid vehicle that may receive a portion of its power from a separately fueled power source, such as an internal combustion engine. For more information, see **J1715_SEP2022** [17].

BBC Dimension

Bumper to back of cab; or a heavy truck, the distance from the front bumper to the rear face of the truck's cab. For more information, see *Motor Truck Engineering Handbook* [7]. See Figure 3.

BCM

Body control module.

BDR

Bendix data recorder.

Bead

The reinforcing part of a tire that is shaped to fit the rim. It is made up of high-tensile-strength steel wrapped and reinforced by the tire sidewall plies (see Figure 7). For more information, see *Motor Truck Engineering Handbook* [7].

FIGURE 7 Cross section of a light vehicle tire showing the bead and other aspects of the tire.

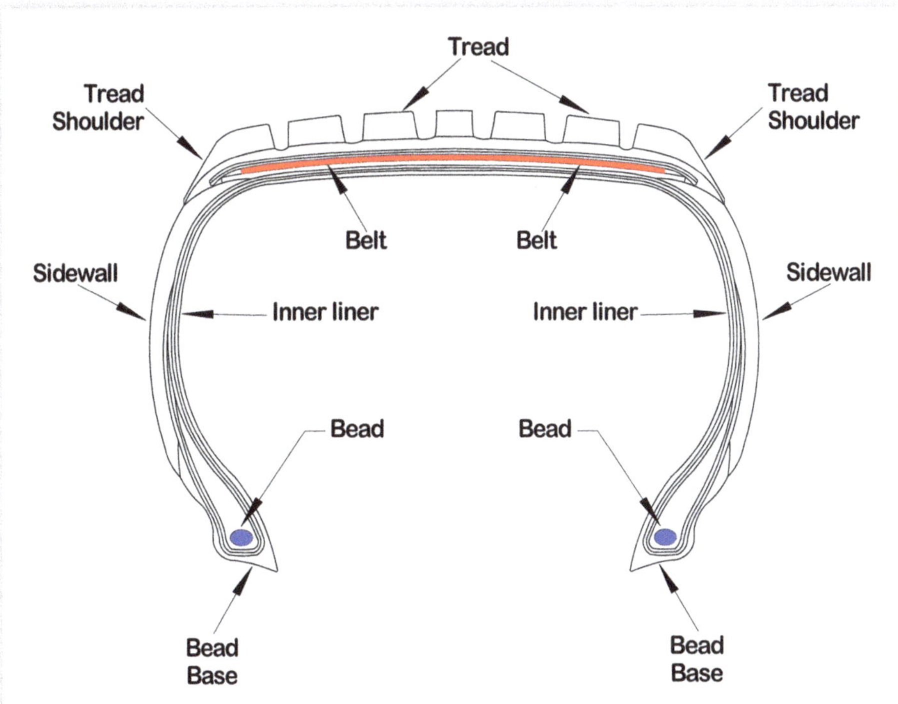

Bead Seat

That part of the rim which provides radial support to the tire and the air pressure seal for tubeless tires. For more information, see **J2047_NOV2019** [15]. See item 21 in Figure 19.

Bead Wire

The wires that make up the bead. Beads come in various types, such as creel or tape, single wound, or cable. They are coated in bronze for adhesion to the rubber. The wires also come in several diameters, the most prominent of which is 0.940 mm (0.037 in.). For more information, see *Tire Forensic Investigation* [26].

Belt

A composite matrix of a reinforcing material (steel wires in most cases, but Kevlar®, nylon, and fiberglass have also been used) and rubber. It will lie on top of the plies in the crown of a tire. The belt will lie at a different angle than the casing plies and will not be connected to the plies, except as they touch near the tire crown. For more information, see *Tire Forensic Investigation* [26].

Belt Edge

When a belt is placed into a tire, each belt has two edges. The belt edges of the working belts normally are the highest stress point of the tire. For more information, see *Tire Forensic Investigation* [26].

Belt Line

A line in a side view that constitutes the lower edge of the design glass outline (DGO) (see Figure 8). For more information, see **J1100_NOV2009** [14].

FIGURE 8 Belt line for a light vehicle.

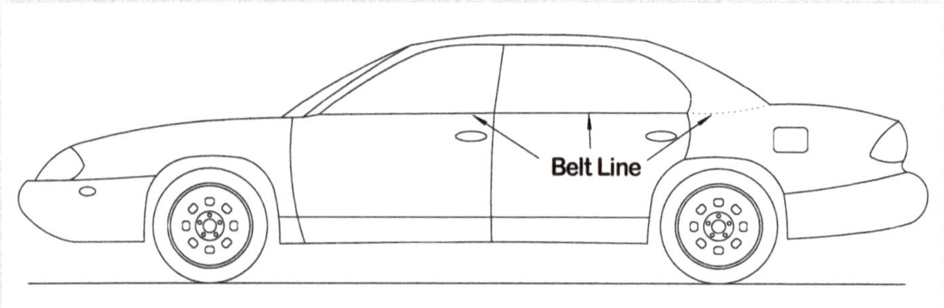

BEV

See Barrier Equivalent Velocity; see also Battery Electric Vehicle.

Bicycle

A vehicle with two or three wheels, solely human-powered and propelled by operable pedals (or a hand-driven equivalent) affixed to a mechanical propulsion system transmitted to one or more wheels, that has handlebars for steering and a saddle-like seat (see Figure 9). A trike or tricycle is a three-wheeled bicycle. For more information, see **J3194_NOV2019** [27].

FIGURE 9 Two-wheeled bicycle.

Vladyslav Starozhylov/Shutterstock.com.

Bicycle Lane

A road lane that has been designated for preferential or exclusive use by bicyclists via pavement markings (and possibly by signs) and is either contiguous with (but marked off from) the traveled way or physically separated from the traveled way. For more information, see **J2944_FEB2023** [28].

Bicycle Model

A two-wheeled vehicle used conceptually in vehicle dynamics studies to represent a four-wheeled vehicle where the side-to-side extent of the vehicle is neglected for simplicity (see Figure 10).

FIGURE 10 Bicycle model depicting many of the parameters used in analysis supported by the model.

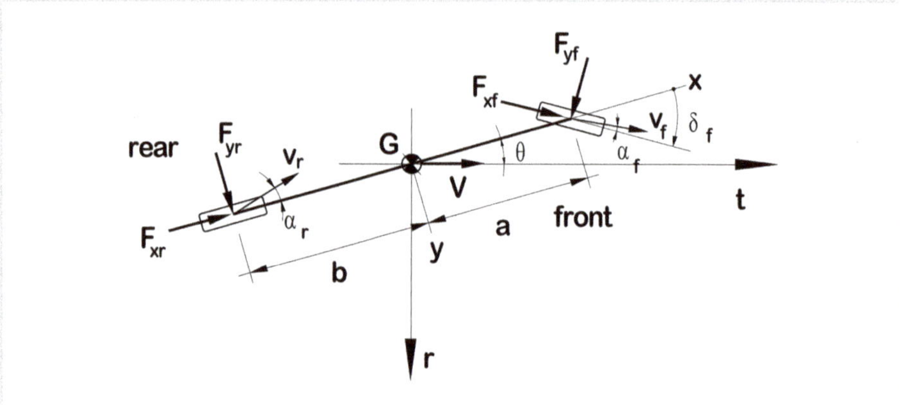

Bight

The intersection of a vehicle seat back and bottom cushion where vehicle belts frequently emerge and where LATCH lower anchorages are placed. In child restraint anchorage systems, the seat bight means the area close to and including the intersection of the surfaces of the vehicle cushion and the seat back. For more information, see **J2939_201304** [2].

Biofidelity

Description of the ability of an anthropomorphic test device (ATD) to respond to crash forces in a manner typical of living human beings, including similar motion patterns, time response, impact responses, acceleration levels, and so on.

Biomechanics

The study of mechanical motion in a biological system. For more information, see *Road Vehicle Dynamics* [8], *Basic Biomechanics of the Musculoskeletal System* [3], *Occupational Biomechanics* [19], *Motor Vehicle Collision Injuries: Biomechanics, Diagnosis, and Management* [20] and *Principles of Biomechanics* [25].

BioRID Dummy

Biofidelic rear impact dummy. A more humanlike dummy than the Hybrid III dummy. For more information, see *Motor Vehicle Collision Injuries: Biomechanics, Diagnosis, and Management* [20].

Bituminous Pavement

A pavement comprising an upper layer or layers of aggregate with a bituminous binder (asphalt, coal tars, natural tars, etc.) and surface treatments such as chip seals, slurry seals, sand seals, and cape seals.

Black Box

Colloquial term used to describe the automobile airbag control module that has the ability to record crash data. See Event Data Recorder.

Blacktop

See Bituminous Pavement.

Blind Spot

Areas surrounding a vehicle that are not directly visible to the driver under a set of circumstances (vehicle geometry, head position and orientation, etc.).

Blind Spot Warning (BSW)

Detects vehicles in the blind spot while driving and notifies the driver to their presence. Some systems provide an additional warning if the driver activates the turn signal. The blind spot is an area to the side and rear of the subject vehicle in an adjacent lane. This notification is most often visual and typically is found on the corresponding A pillar or side mirror. For more information, see **J3063_MAR2021** [23].

BMP

Bitmap. A digital photograph file format.

BMS

Battery management system. Any electronic system that manages a rechargeable battery.

Bobtail

A term used to refer to a truck tractor being operated without a semitrailer.

BOF (BOFRP)

Ball of foot; ball-of-foot reference point. For more information, see **J826_JUN2021** [13] and **J1100NOV2009** [14]. See Figure 26.

Bogie

A combination of two axles usually pivoting about a common trunnion. For more information, see *Motor Truck Engineering Handbook* [7].

Bolt Circle

A circle locating the centers of the bolt holes that are used to attach the wheel to the hub. For more information, see **J2047_NOV2019** [15]. See Figure 19.

Bolt Hole

Mounting stud clearance hole. For more information, see **J2047_NOV2019** [15]. See item 14 in Figure 19.

BOP

Backover protection or backover prevention. A system to prevent or reduce the likelihood of an incident of a vehicle backing over a pedestrian. These systems include backup cameras and proximity detections systems.

Brake

A device whose function is to develop brake torque that directly opposes the longitudinal motion of a vehicle. For more information, see **J670_JUN2022** [6].

Brake Assist System (BAS)

A system that adds additional braking power during emergency braking situations.

Brake Balance

A general term used to describe the distribution of tire braking forces between the axles of a vehicle in relation to the distribution of tire vertical loads at a given steady-state longitudinal acceleration. A "balanced" brake system provides a distribution of tire braking forces that is equal to the distribution of tire vertical loads which prevails as the vehicle decelerates. A "front-biased" brake system will exhibit a higher ratio of front tire braking force to rear tire braking force than the ratio of front tire vertical load to rear tire vertical load. A "rear-biased" brake system has the converse brake balance. For more information, see **J670_JUN2022** [6].

Brake Burnish

The conditioning of a brake's friction surfaces by wear and temperature, either by a test procedure or in-service operation. For more information, see *Motor Truck Engineering Handbook* [7].

Brake Chamber

A component of a pneumatic braking system in which a diaphragm converts air pressure to mechanical force for actuation of a brake (see Figure 11).

FIGURE 11 Double diaphragm brake booster for both service brake operation and parking brake operation.

notsuperstar/Shutterstock.com.

Brake Dive

Pitching of a vehicle due to front suspension compression and rear suspension extension during moderate to heavy braking. For more information, see **J3133_SEP2019** [18].

Brake Drum

The rotating cylinder of a drum brake upon which the brake shoes push during brake actuation (see Figure 12). For more information, see **J670_JUN2022** [6].

FIGURE 12 Brake drums for a light vehicle showing inside (right) and outside (left).

vershandrey/Shutterstock.com.

Brake (Emergency)

Secondary chassis brakes independent in application from the service brakes. Used for controlling a vehicle when the service brake is incapacitated. For more information, see *Motor Truck Engineering Handbook* [7].

Brake (Engine)

Braking device using compression of air by the engine's cylinders, acting through a vehicle's driveline, to decelerate slow the vehicle. This type of brake is frequently referred to as a Jake brake.

Brake Fade

A reduction in braking effectiveness caused by excessive heat and wearing of brake system components.

Brake Pad

The friction material, together with a structural substrate, used in disc brakes. For more information, see **J670_JUN2022** [6].

Brake Pedal Displacement (Brake Pedal Travel)

The linear displacement of the center of the brake pedal relative to its position with zero brake pedal force applied, measured along a chord. For more information, see **J670_JUN2022** [6].

Brake Pedal Force

The force applied to the brake pedal, usually by the driver, tangential to the brake pedal pivot axis. For more information, see **J670_JUN2022** [6].

Brake Rotor

The rotating disc of a disc brake upon which the brake pads push during brake actuation (see Figure 13). For more information, see **J670_JUN2022** [6].

FIGURE 13 Pair of brake rotors for a light vehicle.

Stason4ik/Shutterstock.com.

Brake (Service)

See Service Brake.

Brake Shoe

The friction material, together with a structural substrate, used in drum brakes (see Figure 14). For more information, see **J670_JUN2022** [6].

FIGURE 14 Pair of brake shoes for a heavy truck drum brake.

vershandrey/Shutterstock.com.

Brake Slip
See Wheel Slip.

Braking Distance
The distance taken to bring a vehicle to rest during brake application in straight forward motion. See Stopping Distance.

Braking Force
The force over the contact surface between a tire and a road in the direction of heading of the braked wheel that develops as a result of brake application.

Braking Force (Peak)
The largest force that can be developed during brake application as wheel slip is varied over the range of free-rolling slip to locked-wheel slip.

Breakdown
The failure of the dielectric of an insulation due to the effects of an excessive electric field and/or physical or chemical deterioration of the dielectric material. For more information, see **J1715_SEP2022** [17].

Bridge Formula
A formula used to determine the maximum gross weight permissible on any group of axles. For more information, see *Motor Truck Engineering Handbook* [7].

BSD
Blind spot detection. A system that monitors behind and adjacent to a vehicle and warns the driver of the presence of objects in these areas.

BSM
Blind spot monitoring.

BSW
See Blind Spot Warning.

BTO
Brake throttle override. A vehicle system that retards the engine throttle if both the throttle and the brake are actuated simultaneously.

BTS

Bureau of Transportation Statistics; part of the US Department of Transportation. The bureau is a source of statistics on commercial aviation, multimodal freight activity, and transportation economics, and it provides context to decision makers and the public for understanding statistics on transportation.

BTSI

Brake transmission shift interlock. A system that prevents the transmission gear selector lever from shifting out of park when the brake pedal is not actuated.

BTU

British thermal unit; a measure of heat with units of energy.

Buckle

The primary quick release component of a seat belt assembly that accepts and connects to a latch plate and is intended to facilitate wearing and removal of the seat belt by a vehicle occupant (see **Figure 15**). For more information, see **J1803_JAN2013** [16].

FIGURE 15 Seat belt buckle.

Bus

A vehicle designed to transport more than 15 passengers, including the driver. See also Motorcoach.

Buttress

The exterior surface part of the tire between the upper sidewall and the tread surface edge. For more information, see *Tire Forensic Investigation* [26].

C

CA

Cab-to-axle; the distance from back of cab to centerline of rear axle. For more information, see *Motor Truck Engineering Handbook* [7]. See Figure 3.

CAA

Clean Air Act. The Clean Air Act (42 U.S.C. 7401 et seq.) is a comprehensive federal law that regulates all sources of air emissions.

CACC

Cooperative adaptive cruise control.

CAD

Computer-aided design. Often refers to drafting software.

Cadence

In walking, the number of steps per unit of time. In bicycling, the rate of pedal rotation per unit of time. For more information, see *Principles of Biomechanics* [25].

CAFE

Corporate Average Fuel Economy.

CAN

Controller Area Network, A type of communication bus (see also GMLAN).

Camber

Amount of inward or outward tilt in the vertical axis of a wheel. Camber is positive when the top of the wheel is rotated away from a vehicle's centerline and negative when the top of the wheel is rotated toward the vehicle centerline (see Figure 16).

FIGURE 16 Diagram showing positive and negative camber of the front wheels.

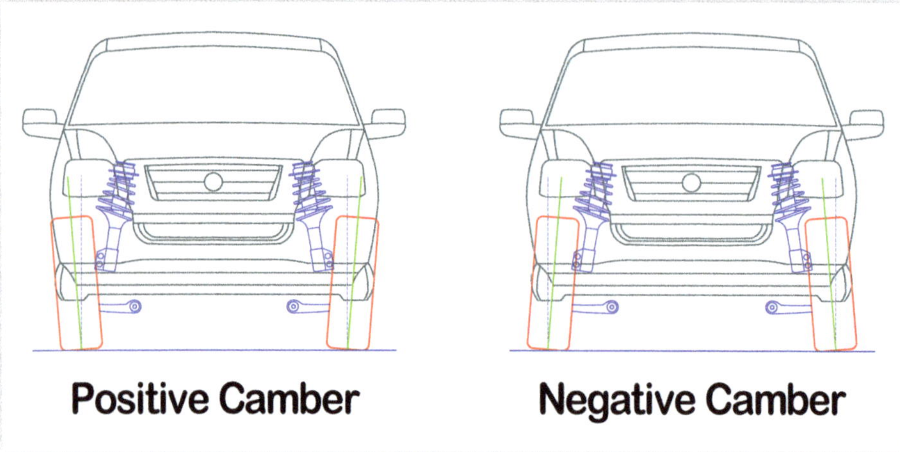

Camber Force (Camber Thrust)

The lateral force when the slip angle is zero and the ply steer and conicity forces have been subtracted. For more information, see *Road Vehicle Dynamics* [8].

CARMA

Cooperative Automation Research Mobility Applications.

Car Seat

A common alternative term for child restraint system, especially when speaking with general audiences. For more information, see **J2939_201304** [2].

CAS

Collision avoidance system.

Caster Angle

The angle in a side elevation between the steering axis and the vertical. It is considered positive when the steering axis is inclined rearward (in the upward direction) and negative when the steering axis is inclined forward (see Figure 17). For more information, see *Road Vehicle Dynamics* [8].

FIGURE 17 Positive and negative caster angles.

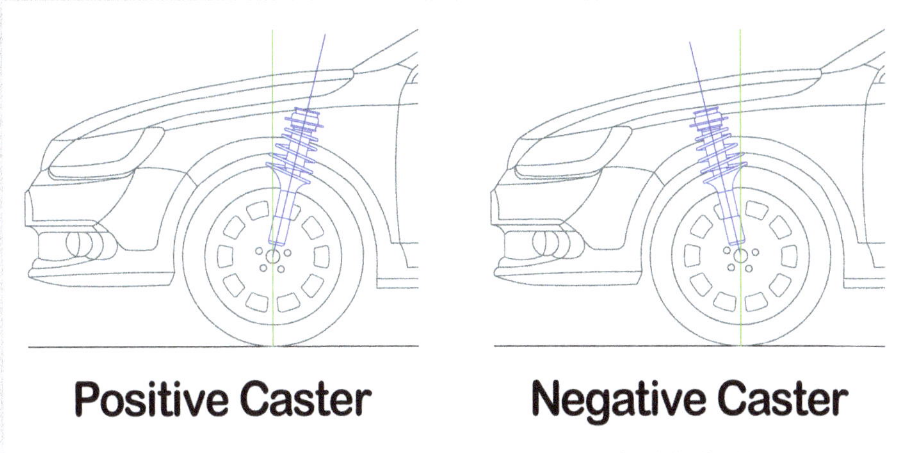

Caster Offset

The distance in a side elevation between the point where the steering axis intersects the ground and the center of tire contact. The offset is considered positive when the intersection point is forward of the tire contact center and negative when it is rearward. For more information, see *Road Vehicle Dynamics* [8].

CCC

See Conventional Cruise Control.

CCD

Charge-coupled device; and example is the sensor in a digital camera.

CCM

Cruise control module.

CDA

See Cooperative Driving Automation.

CDL

Commercial driver's license.

CDR

Crash Data Retrieval. A system of software and hardware, sold by Bosch Automotive Service Solutions, Inc., used to image electronic event data from certain light vehicles.

CDW

Collision detection warning.

Center Hole

On a wheel, the clearance hole for the pilot of the hub. For more information, see **J2047_NOV2019** [15]. See item 5 in Figure 19.

Center of Gravity (CG)

That point of a body through which the resultant force of gravity (weight) acts irrespective of the orientation of the body.

Center of Impact

See Impact Center.

Center of Mass

See Center of Gravity.

Central Impact

An impact in which the contact impulse passes through the center of gravity. See Oblique Impact.

Centripetal Acceleration

The scalar value of the component of vehicle acceleration in the direction of the horizontal normal to the horizontal velocity vector. For more information, see **J670_JUN2022** [6].

Cervical

Pertaining to the neck. For more information, see **J885_FEB2011** [4].

CF

Cab-to-frame; for a heavy truck, the distance from the back of the cab to the rear end of the frame. See Figure 3.

CFR

Code of Federal Regulations.

Change in Momentum

(1) The difference of the momentum (product of mass and velocity) of a mass from one time to another; (2) the difference of the momentum of a mass between the beginning and end of contact with another mass; (3) the difference in the momentum of a system of bodies. See also Conservation of Momentum.

Change of Velocity

The difference between velocity vectors at two points in time; represented as ΔV. See Delta-V.

Channel

Refers to the number of modulators controlled by a vehicle's ABS electronic control unit (ECU). A two-sensor/one modulator valve system would be considered a single-channel system. A 2S/2M system is a two-channel system.

Charge Sustaining (CS) Mode

An operating mode where the HEV runs by consuming fuel while sustaining the electric energy of the rechargeable energy storage system (RESS). For more information, see **J1715_SEP2022** [17].

Child Restraint Anchorage System

A standardized system of user-ready hardware in vehicles, popularly referred to as LATCH in the USA (similar to the Canadian lower Universal Anchorage System or LUAS), consisting of two lower anchorages and one upper anchorage specified in FMVSS 225, for installing child restraint systems independent of the vehicle's seat belt. For more information, see **J2939_201304** [2].

Child Restraint System (CRS)

A general term used in FMVSS 213 for devices designed "to restrain, seat, or position children who weigh 80 lbs or less." These include rear-facing restraints (infant-only and convertible), forward-facing restraints (convertible, child seat, combination seat), car beds, harnesses, and boosters (belt-positioning and shield). For more information, see **J2939_201304** [2].

Chop

(1) A broad shallow gouge in a road surface, beginning with an even, regular, deeper side and terminating in scratches and striations on the opposite

shallower side; (2) a depression in pavement made by a strong, sharp metal edge moving under heavy pressure more commonly occurring at an impact event as opposed to postimpact trajectory. For more information, see *Traffic Accident Investigation Manual* [29].

Chop the Throttle

For a motorcycle, a common phrase for rapidly closing the throttle from an open position. For more information, see **J3133_SEP2019** [18].

CIB

Crash imminent braking or collision imminent braking.

CIREN

Crash Injury Research and Engineering Network. A program run by the National Highway Traffic Safety Administration (NHTSA) that utilizes medical and engineering centers to improve the prevention, treatment, and rehabilitation of injuries from motor vehicle crashes.

Class I Electrical System

An electrical system with basic insulation throughout, whose conductive accessible parts are connected to the protective earthing conductor and provided with an earthing terminal or connection to a vehicle. For more information, see **J2578_AUG2014** [30].

Class II Electrical System

An electrical system with double insulation and/or reinforced insulation throughout. For more information, see **J2578_AUG2014** [30].

Clearance Lamp

Light used on the front or rear of a motor vehicle to indicate overall width or height.

Closing Speed

(1) The magnitude of the relative velocity between two vehicles at a given point in time as they approach each other; (2) the relative velocity between two vehicles as they approach each other at the beginning of an accident; (3) a normal component of the closing velocity. See Approach Speed.

Closing Velocity

(1) The magnitude of the relative velocity between two vehicles at a given point in time as they approach each other; (2) the magnitude of the relative velocity between two vehicles at the beginning of a crash; (3) the vector difference between the velocity of the vehicle and the vehicle/object struck immediately before impact.

CMB

Collision mitigation braking.

CMS

Camera monitor system. Also collision mitigation system.

CMV

Commercial motor vehicle.

CO

Carbon monoxide.

CO2 (CO_2)

Carbon dioxide.

Coefficient of Friction

A number representing the resistance to sliding of two flat surfaces in contact; defined as the ratio of the resistance force to the normal force between the surfaces. See Frictional Drag Factor. As defined for a tire and wheel assembly, the ratio of the maximum value of the tangential force between the tread rubber and the horizontal road surface to the absolute value of normal force attainable on a given road surface on a given rolling or locked wheel at specified test conditions. For more information, see **J2047_NOV2019** [15].

Coefficient of Restitution

The ratio of the relative normal velocity at the time of separation to the relative normal velocity at the time of initial contact between the point area of contact of two colliding bodies. In the field of crash reconstruction, the coefficient of restitution is an empirical parameter.

Coefficient of Rolling Resistance

The ratio of the force of resistance to rolling with zero slip to the vertical load of a wheel or vehicle. See Rolling Resistance.

Cognitive Distraction

Any time duration of driver cognitive loading that competes with activities necessary for safe driving. For more information, see **J3077_DEC2015** [24].

Collinear Collision

A collision between two vehicles in which their respective directions of travel are parallel to one another, either as a rear-end or head-on collision. For more information, see *Road Vehicle Dynamics* [8].

Collision

Sudden contact of a vehicle with an object or another vehicle, usually resulting in the change in momentum of at least one of the objects and generally includes visible damage to one or both vehicles/objects. See Crash; Impact.

Collision Deformation Classification (CDC)

A classification of the extent of deformation to an automobile, utility vehicle, pickup, and van from a crash. For more information, see **J224_MAY2022** [31].

Collision Intervention

Collision intervention systems provide momentary intervention during potentially hazardous situations. Depending on the situation, the collision intervention system action may result in collision avoidance, or it may result in a reduction in the severity of the collision. Due to the momentary nature of the actions of these systems, their intervention does not change or eliminate the role of the driver. For more information, see **J3063_MAR2021** [23].

Combined Braking System (CBS)

Where at least two brakes on different axles are actuated by the operation of a single control. For more information, see **J3133_SEP2019** [18].

Common Contact Point

See Impact Center.

Common Velocity Conditions

Two independent conditions applicable to a collision where at the time of separation the relative normal velocity component is zero (no restitution) and the relative tangential component of velocity is zero (sliding has ended).

Compatibility (Vehicle)

Disparities in structural crashworthiness of different-sized vehicles due to varying structural geometries, such as different heights of the vehicles' front and side structures; a tendency of some vehicles to inflict more damage on another vehicle in a crash.

Competing Activity

An activity or activities which place/s demands on cognitive, auditory, vocal/verbal, visual, motoric, and other resources separately or in any combination with demands that are the same as or similar to the resources demanded by safe driving (hence giving rise to resource competition), and which occur concurrently while driving. For more information, see **J3077_DEC2015** [24].

Complacency

In an automated driving system, overconfidence on the part of a driver or fallback-ready user in his or her knowledge of what to expect in terms of performance by a given level 1 through 2 driving automation feature; such overconfidence may lead to overreliance on the system. For more information, see **J3114_DEC2016** [32].

Concentric Work

In biomechanics, work produced by a muscle when it contracts and shortens. For more information, see *Basic Biomechanics of the Musculoskeletal System* [3].

Concrete Pavement

A solidified pavement made by mixing cement and an aggregate composed of hard, inert particles of varying size, such as a combination of sand or broken-stone screenings, with gravel, broken stone, lightweight aggregate or other material. For more information, see *Marks' Standard Handbook for Mechanical Engineers* [33].

Concussion

Loss of consciousness. For more information, see *Principles of Biomechanics* [25].

Conservation of Momentum

The principle of physics for vehicles in dynamic contact stating that, in the absence of external forces, the sum of the preimpact momentum is equal to the sum of the postimpact momentum of the vehicles.

Conspicuity

The quality of being conspicuous (e.g., an object that clearly contrasts with its background).

Contact Damage

Deformation sustained in a vehicle from physical engagement with another vehicle or object. See Dynamic Crush, Induced Damage; Residual Crush.

Contact Patch

The area or region of mutual contact between a tire and the surface over which it rests or moves.

Contact Point

The point of intersection of the resultant contact impulse with the intervehicular contact surface of each of two colliding vehicles. See Impact Center.

Contact Surface

See Intervehicular Contact Surface.

Contrast

The difference in luminance (lightness or brightness) between two areas. Providing sufficient contrast is the most important element for ensuring pattern recognition (i.e., being able to read a display or recognize a symbol). For more information, see **J264_OCT2021** [34].

Control Arm (Wishbone)

A suspension part generally used to connect a knuckle or axle to the sprung mass (see Figure 18). For more information, see **J670_June 2022** [6].

FIGURE 18 Lower (wishbone) control arm.

Sharomka/Shutterstock.com.

Contusion

Bruising from a direct impact. For more information, see **J885_FEB2011** [4].

Conventional Cruise Control (CCC)

System capable of controlling the speed of a vehicle, as selected by the driver without consideration of in-path forward vehicles. For more information, see **J2399_OCT2021** [10].

Conventional Vehicle

A vehicle designed to be operated by a conventional driver during part or all of every trip. A conventional vehicle may be equipped with one or more level 1 or level 2 driving automation system features that support the driver in performing the dynamic driving task (DDT), but do not perform the complete DDT, as well as level 3 and level 4 automated driving system (ADS) features that require a conventional driver to operate the vehicle during portions of each trip. For more information, see **J3016_JUN2018** [9].

Converter Dolly

A device that consists of one or two axles and a fifth wheel to which a semitrailer can be attached to form a trailer. Generally used to allow for towing two or three trailers behind a single truck tractor, referred to as doubles or triples, respectively.

Convertible Child Restraint

A child restraint that can be used rear-facing for infants up to at least 22 lb or as much as 35 lb, and then turned to face forward until the child reaches the product's upper weight limit, usually at least 40 lb to 65 lb. Most current convertible restraints can accommodate children rear-facing up to 40 lb. For more information, see **J2939_201304** [2].

Convulsion

Violent erratic muscle contraction. For more information, see *Principles of Biomechanics* [25].

Cooperative Driving Automation (CDA)

Automation that uses M2M communication to enable cooperation among two or more entities with capable communications technology and is intended to facilitate the safer, more efficient movement of road users, including enhancing performance of the dynamic driving task (DDT) for a vehicle with driving automation feature(s) engaged. For more information, see **J3216_JUL2021** [11].

Coordinate System, Vehicle

See Three-Axis Vehicle Coordinate System.

Cornering Coefficient

See Sideslip Coefficient.

Cornering Stiffness

The absolute value of the first derivative of lateral force with respect to slip angle, usually determined at zero slip angle, zero inclination angle, zero longitudinal slip, and zero path curvature; typically designates C_α. For more information, see **J670_JUN2022** [6] and **J2047_NOV2019** [15]. See Sideslip Coefficient.

Counter Steer (Initiating a Turn)

For a motorcycle, when a rider inputs a steer torque in the direction opposite to that which the vehicle is turning or is intended to turn. The counter steering torque necessary to initiate the roll angle required for a given turn should be distinguished from the sustained steer torque and steering angle necessary to maintain a constant radius and roll angle in a steady-state turn. At all but the lowest speeds, the initial counter steer torque and angle are both opposite the desired turn direction. The sustained trim steer angle is typically in the same direction as the turn. For more information, see **J3133_SEP2019** [18].

Cover the Brake

A condition in which the driver of a vehicle (or rider of a motorcycle) maintains a position near, or in light contact with, the brake control to decrease reaction time in applying the brakes for an emergency stop. This may apply to either the front or rear brake controls on a motorcycle. For more information, see **J3133_SEP2019** [18].

CPR

Crash pulse recorder. A device that measures acceleration during a crash. See Accelerometer.

Crack Pressure

In a vehicle with air brakes, that signal pressure when applied to a brake valve will cause the piston or diaphragm to move such as to deliver output pressure. For more information, see *Commercial Vehicle Braking Systems: Air Brakes, ABS and Beyond* [5].

CRAF

Civil Reserve Air Fleet.

Crankcase Ventilation

Vent system for a crankcase that provides for burning of oil vapor rather than exhausting it to the atmosphere. For more information, see *Motor Truck Engineering Handbook* [7].

Crash

An event in which one or more vehicles make unintended contact with another vehicle or other object, producing injury, death, and/or property damage. See Accident; Collision; Impact.

Crash Bars

On motorcycles, crash bars, more appropriately called engine guards, are intended to protect the engine and body panels from contact with the roadway. A common arrangement is a loop of steel tubing mounted to each side of a motorcycle's lower frame.

Crash Duration
The period of time defined by the moment when two vehicles come in contact until that time when they separate.

Crash Pulse
The shape of the intervehicular force curve during the crash duration.

Crash Reconstruction
See Accident Reconstruction.

Crash Scene
The area and environment where a crash occurred, including the area of impact, areas of rest, and locations of physical evidence while the vehicles and the people involved, including emergency responders, are still present.

Crash Site
The area and environment where a crash occurred, including the area of impact, areas of rest, and locations of physical evidence after the vehicles and the people involved have moved out of the area.

CRASH3
An acronym for **C**ALSPAN **R**econstruction of **A**ccident **S**peeds on the **H**ighway, Version 3; a method of reconstruction that uses the calculation of the crush energy of a collision and an approximate postimpact trajectory spinout simulation.

Crashworthiness
The ability of a motor vehicle to withstand forces and energies occurring in a crash and to protect its occupants from injuries due to those forces and energies.

Critical Speed
The maximum speed at which a vehicle can traverse a path with a specific radius of curvature without loss of directional control; the speed of a vehicle undergoing a turn maneuver at which the tires leave visible sideslip marks. The speed when the centripetal acceleration is limited by roadway friction: $f = \mu\, n \pm m$, where f is the drag factor, μ is the coefficient of friction, n is the percentage of braking, and m is the slope. This is valid for slopes less than 10°. For more information, see **J2969_201701** [35].

Critical Speed Formula

A formula, $v_{cr} = \sqrt{fgR}$, that calculates the speed of a vehicle from its radius of curvature, R, frictional drag coefficient, f, and acceleration of gravity, g. For more information, see *Vehicle Accident Analysis and Reconstruction Methods* [1].

A formula that calculates the speed of a vehicle from the radius of curvature of its tires critical speed scuff marks, taking into account roadway grade, superelevation and frictional drag coefficient. For more information, see **J2969_201701** [35].

Critical Speed Scuff

A tire mark left by a rotating wheel that is sliding or slipping sideways due to the centrifugal force exceeding the peak lateral friction. The skid mark normally will have striation marks. Also referred to as centrifugal skid or inertial mark. For more information, see *Road Vehicle Dynamics* [8].

Crown

The lateral slope of a roadway where the center is higher than the outside edge, usually for drainage purposes. For more information, see *Road Vehicle Dynamics* [8].

Cruise Control

A system on a vehicle that automatically maintains a selected speed setting without the driver maintaining pressure on the accelerator pedal. See Adaptive Cruise Control.

Crumple Zone

That portion of the front or rear of a vehicle designed to absorb the energy of a collision for the protection of the occupants.

Crush Area

Area defined by the original vehicle exterior and a crush profile.

Crush Equivalent Speed

See Energy Equivalent Speed (EES).

Crush Profile

The geometric shape in a specified plane (e.g., vertical, horizontal) that describes the vehicle damage resulting from an impact.

Crush Stiffness

See Crush Stiffness Coefficient.

Crush Stiffness Coefficient

An empirical quantity used in the calculation of the energy dissipated in a collision and associated with each vehicle's velocity change, ΔV. See CRASH3.

CTA

Cross-traffic alert. An automatic warning system, typically at the rear of a vehicle, that alerts a driver to approaching traffic when backing out of a parking place or backing out onto a roadway. Sometimes referred to as a rear cross-traffic warning system, or RCTW.

CTE

Coefficient of thermal expansion. A numerical value that characterizes how the size of an object changes with a change in temperature.

Curb Trip

A rollover that occurs when a vehicle moving laterally slides into a length of raised curb; the rollover is initiated by the impact of the wheels with a raised curb. For more information, see **J2926_201707** [36].

Curb Weight

The weight of the base vehicle (standard equipment only), with all fluids filled to maximum (fuel, oil, transmission, coolant, etc.). For heavy trucks, the curb weight does not include engine fuel. For more information, see **J1100_NOV2009** [14].

Cutaneous

Referring to the skin. For more information, see *Principles of Biomechanics* [25].

CVT

Continuously variable transmission. A transmission that changes through continuous gear ratios, typically by positioning of a flexible drive belt along the conical faces of the driving and driven pulleys as a vehicle is moving, rather than through a fixed set of gear ratios that is characteristic of a mechanical transmission. For more information, see *Bosch Automotive Handbook* [37].

D

DA

Detroit Assurance.

Damper

A suspension part used to provide damping, most commonly in bounce and roll. A typical damper provides primarily viscous damping, with some attendant Coulomb friction. For more information, see **J670_JUN2022** [6]. See also Shock Absorber; Strut.

DCFC

Direct current fast charger. A battery charging system that uses a three-phase 480 volt alternating current (AC) electric circuit but delivers direct current (DC) to a vehicle.

DDD

Driver drowsiness detection.

DDT

See Dynamic Driving Task.

DDW

Drowsy driver warning.

DFW

Driver fatigue warning.

Debeaded (Unbuttoned)

The position of the tire versus its position on the rim bead seat. A tire that is unbuttoned from the rim is physically on the rim, but the bead or beads are no longer seated; therefore, the tire cannot hold air. A tire that is buttoned on the rim can hold air because the beads are seated properly. For more information, see *Tire Forensic Investigation* [26].

Decibel

A logarithmic measure of the level, L, of a time-varying signal, $s(\tau)$, relative to a reference value s_{ref}. $L = 10\log\frac{s^2}{s_{ref}^2}$ where s^2 is the mean square value of the signal.

Degloving

The traumatic tearing away of skin and flesh from the bones. For more information, see *Principles of Biomechanics* [25].

Delta-t (Δt or $\Delta\tau$)

A time interval associated with an event such as vehicle-to-vehicle contact; the time duration of impulse.

Delta-V (ΔV)

The difference or change of a velocity vector over a time interval; the difference in the velocity vector of the center of gravity of a vehicle between separation and first contact in a crash.

Departure Velocity

See Separation Velocity.

Deployment (Event)

Actuation of a supplementary restraint, based on an enabling algorithm, following which acceleration and other data are recorded and made available to an event data recorder.

Deployment Level (Event)

An acceleration level sufficient to cause the General Motors Sensing and Diagnostic Model's (SDM) crash-sensing algorithm to "enable" and anticipate a collision severity which otherwise warrants a deployment for that vehicle, but a deployment had been previously commanded.

Design Glass Outline (DGO)

An outline on the surface of a vehicle glass assembly which defines the boundary of transparent area. It can be used to determine certain vision angle measurements. See Belt Line and Figure 8. For more information, see **J1100_NOV2009** [14].

Designated Seating Position

Any seating location that provides a seat surface width (SSW) of at least 330 mm. For more information, see **J1100_NOV2009** [14].

DGO

See Design Glass Outline.

DGPS

Differential global positioning system.

Diagnostic Trouble Code (DTC)

Information provided by an on-board diagnostic system for operators and repair technicians.

Direction of Principal Force (DOPF)

See Principal Direction of Force (PDOF).

Disc

That part of the wheel which is the supporting member between the hub (axle) and the rim (see item 2 in Figure 19). For more information, see **J2047_NOV2019** [15].

FIGURE 19 Disc wheel.

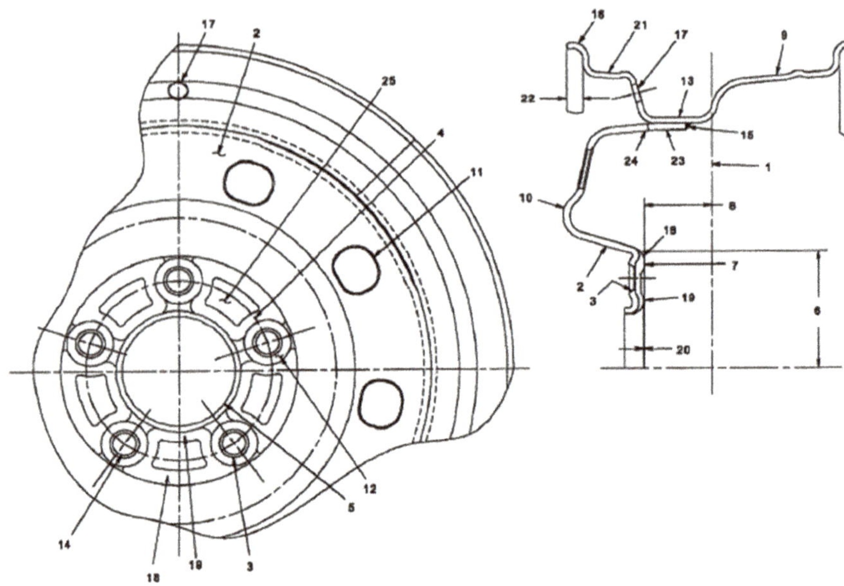

1. Rim Center Plane
2. Disc
3. Nut Seat
4. Pitch Circle Diameter of Bolt Holes
5. Center Hole Diameter
6. (See Figure 3)
7. Attachment Face
8. Inset
9. Rim
10. Hat
11. Window
12. Nut Boss
13. Well
14. Bolt Hole
15. Assembly Weld
16. Rim Flange
17. Valve Hole
18. Outer Mounting Pad
19. Inner Mounting Pad
20. Step
21. Bead Seat
22. Flange Offset
23. Disc Flange
24. Disc Scallop
25. Rib

Dispatch (in Driverless Operation)

To place an ADS-equipped vehicle into service in driverless operation by engaging the automated driving system. The term "dispatch" as used outside of the context of ADS-equipped vehicles, is generally understood to mean sending a particular vehicle to a particular pick-up or drop-off location for purposes of providing a transportation service. In the context of ADS-equipped vehicles this term includes software-enabled dispatch of multiple ADS-equipped vehicles in driverless operation that may complete multiple trips involving pickup and drop-off of passengers or goods throughout a day or other predefined period of service, and which may involve multiple agents performing various tasks related to the dispatch function. For more information, see **J3016_JUN2018** [9].

Disposal

The deployment command of the second (or higher, if present) stage of a frontal airbag for the purpose of disposing of the propellant from the airbag device.

Distal

In a limb: further away from the limb's attachment to the body. Elsewhere: further away from the central axis of the body. For more information, see *Occupational Biomechanics* [19].

Divot

A piece of turf or sod torn up by dynamic contact.

DLC

Diagnostic link connector. This may also be known as a data link connector.

DMS

Driver monitoring system. A vehicle safety system that assesses the alertness of the driver of a vehicle.

DOE

Department of Energy. The mission of the Energy Department is to ensure America's security and prosperity by addressing its energy, environmental, and nuclear challenges through transformative science and technology solutions.

DOI

Department of the Interior. The US department that protects and manages the nation's natural resources and cultural heritage.

Dolly

Two-wheel trailer equipped with a drawbar and the lower portion of a fifth wheel and other components necessary to permit a semitrailer and dolly combination to operate as a full trailer; sometimes called a "PUP." For more information, see *Motor Truck Engineering Handbook* [7].

Dolly Hitch

Coupling device between the leading trailer and dolly consisting of a pintle hook and drawbar connection. For more information, see *Motor Truck Engineering Handbook* [7].

Dorsiflexion

Bending about the ankle and wrist joints in a dorsal direction. For more information, see *Basic Biomechanics of the Musculoskeletal System* [3].

DoT, DOT

Department of Transportation.

Double Cap Nut

The inner and outer nuts used to secure stud-mount wheels to a vehicle. The inner dual wheel is attached by an inner cap nut with a spherical radius and the outer dual wheel is attached by an outer cap nut with a spherical radius. For more information, see *Motor Truck Engineering Handbook* [7].

Double Clutching

A technique used in shifting a manual transmission, especially one without synchronization, in which the driver initially disengages the clutch while shifting out of the current gear, then reengages the clutch to match engine speed with transmission speed for the desired gear, then again disengages the clutch while selecting the desired gear, and finally reengages the clutch to proceed in the desired gear.

Double Insulation

A system of two independent insulations, each of which is capable of acting as the sole insulation between live and accessible parts in the event of failure of the other insulation. The insulation system resulting from a combination of basic and supplementary insulation. For more information, see **J2578_AUG2014** [30].

Doubles (Double Bottom)

Truck tractor with a semitrailer and full trailer combination. For more information, see *Motor Truck Engineering Handbook* [7].

D-Point

A point on the bottom surface of the HPM-II cushion pan, at the lateral centerline, 25.5 mm (15°) rearward of the H-point. On the HPM, the D-point is the lowest point on the centerline of the bottom of the cushion pan in the installed position. See Figure 26. For more information, see J826_JUN2021 [13] and **J1100NOV2009** [14].

Drag Factor

An equivalent acceleration expressed as a fraction of the acceleration of gravity, *g*. Also see Frictional Drag Coefficient.

Drag Force

Negative tractive force. Generally, for a vehicle, this force acts at the tire-ground contact patches and is generated by application of the vehicle's service brakes.

Drag Link

The mechanical link between the Pitman arm on the steering gear and the steering arm on a front (steer) axle.

Drag Sled

A weighted device (whose bottom surface is covered with a portion of tire tread) that is pulled along a roadway surface and provides a sliding friction coefficient of that device and roadway surface by computing the ratio of the pull force to its weight.

Drift

The limit condition for vehicle directional response wherein the front and rear tires have both reached their cornering limit. For more information, see **J670_JUN2022** [6]. See Plow; Spin.

Driver

In the context of automated driving systems, a user who performs in real time part or all of the dynamic driving task (DDT) and/or DDT fallback for a particular vehicle. In a vehicle equipped with a driving automation system, a driver may assume or resume performance of part or all of the DDT from the driving automation system during a given trip. For more information, see **J3016_JUN2018** [9].

Driver Automation Abuse

The intentional misuse of a driving automation system feature in a manner contrary to the manufacturer's instructions. For more information, see **J3114_DEC2016** [32].

Driver Distraction

The diversion of attention away from activities critical for safe driving toward a competing activity, which may result in insufficient or no attention to activities critical for safe driving. For more information, see J3077_DEC2015 [24]; Mismatch between the driver's current resource allocation and that demanded by activities critical for safe driving, rather than in terms of attentional failures of the driver. For more information, see **J3198_202010** [38].

Driver Workload

The amount of physical and mental activity that is required to perform a particular task or set of tasks while driving. For more information, see **J3077_201512** [24]. See Average Driver Workload; Instantaneous Driver Workload; Overall Driver Workload.

Driverless Operation

Operation of an automated driving system–equipped vehicle in which either no onboard user is present or in which onboard users are not drivers or fallback-ready users. For more information, see **J3016_JUN2018** [9].

Driveshaft

An assembly of two universal joints and a solid tubular member typically used to transmit engine power to a vehicle's drive axle.

Drivetrain

Clutch, transmission, driveshaft, and drive axles. For more information, see *Motor Truck Engineering Handbook* [7].

Driving Automation

The performance by hardware/software systems of part or all of the DDT on a sustained basis. For more information, see **J3016_JUN2018** [9] and **J3216_202107** [11].

Driving Automation Bias

The omission and commission errors that result from the use of driving automation system–issued cues as a replacement for vigilant information seeking and processing. For more information, see **J3114_201612** [32].

Driving Automation Disuse

The intentional nonuse of a driving automation system feature. For more information, see **J3114_DEC2016** [32].

Driving Automation Misuse

The unintentional use of a driving automation system feature in a manner contrary to the manufacturer's instructions. For more information, see **J3114_DEC2016** [32].

Drop Axle

An unpowered auxiliary axle on a truck, truck tractor, or trailer that can be raised or lowered to change the vertical load distribution of permanent axles. This axle is called a "pusher" if it is positioned ahead of the powered tandem axles, or a tag axle if it is positioned behind the powered tandem axles. A drop axle can also refer to an axle, for example, the steer axle on a heavy truck, that has its center section "dropped" below the height of the axle's wheel spindles.

Drowsiness

The inclination to sleep resulting from lack of sleep, boredom, hunger, or other outside factors. For more information, see **J3198_202010** [38].

DTC

See Diagnostic Trouble Code.

Dual-Mode Vehicle

A type of automated driving system–equipped vehicle designed for both driverless operation and operation by a conventional driver for complete trips. For more information, see **J3016_201806** [9].

Dual Wheels

A set of two closely spaced wheels on each end of an axle, typically used on heavy trucks, truck tractors, and trailers (see Figure 20).

FIGURE 20 Dual wheels on tandem axles on a heavy truck.

J.J. Gouin/Shutterstock.com.

Dump Body

A truck or trailer body generally hinged at the rear, used to carry loose, bulk materials, and emptied by raising the forward end of the body, typically with a hydraulic actuator. Its capacity is usually given in cubic yards.

Dump the Clutch

Common motorcyclist term to describe an abrupt release of the clutch lever that results in immediate engagement of the engine to the drivetrain. At lower engine speeds, this action may cause the motorcycle to lurch forward and/or stall the engine. At higher engine speeds, this action may cause the rear wheel to spin or the motorcycle's front wheel to lift off the ground. For more information, see **J3133_201909** [18].

Dura Mater

Thick outermost layer of the brain. For more information, see *Principles of Biomechanics* [25].

DVI
Driver-vehicle interface.

DWPT
Dynamic wireless power transfer. A battery charging system that utilizes devices embedded in roadways to charge a vehicle's batteries as the vehicle travels over the device or it stops or parks over the device.

DWT
Deadweight tonnage. In the marine shipping industry, the amount of weight that a ship can carry.

DXF
Drawing Exchange Format. A graphical format for drawings made with computer-assisted design (CAD) programs.

Dynamic Crush
The deformation formed by the external surface of a vehicle at any time during an impact, usually measured relative to the corresponding as-manufactured undeformed surface. See Crush Area; Crush Profile; Residual Crush; Static Crush.

Dynamic Driving Task (DDT)
All of the real-time operational and tactical functions required to operate a vehicle in on-road traffic, excluding the strategic functions, such as trip scheduling and selection of destinations and waypoints, that include the following:

- Lateral vehicle motion control via steering (operational)
- Longitudinal vehicle motion control via acceleration and deceleration (operational)
- Monitoring the driving environment via object and event detection, recognition, classification, and response preparation (operational and tactical)
- Object and event response execution (operational and tactical)
- Maneuver planning (tactical)
- Enhancing conspicuity via lighting, signaling and gesturing, etc. (tactical)

For more information, see **J3016_201806** [9].

Dynamic Driving Task Fallback (DDT Fallback)

The response by the user to either perform the DDT or achieve a minimal risk condition after occurrence of a DDT performance-relevant system failure(s) or upon operational design domain (chop) exit, or the response by an automated driving system (ADS) to achieve minimal risk condition, given the same circumstances. For more information, see **J3016_201806** [9].

E

EBS (Electronic Braking System)

A braking system that utilizes electronics to replace the pneumatic logic of an air-actuated commercial vehicle braking system. It includes an automatic braking system (ABS) and automatic traction control (ATC) as integrated functions. EBS is sometimes referred to as "brake by wire." For more information, see *Commercial Vehicle Braking Systems: Air Brakes, ABS and Beyond* [5].

Eccentric Contraction

Increase of tension within a muscle due to resisting an applied load while lengthening. For more information, see *Occupational Biomechanics* [19].

Eccentric Impact

See Oblique Impact.

Eccentric Work

In biomechanics, work produced by a muscle when it lengthens while resisting an applied load.

ECE

Economic Commission for Europe. One of five regional commissions under the jurisdiction of the United Nations Economic and Social Council.

ECI

Electronic conductive immunity.

ECM

Engine control module. An automotive ECM, also called an electronic control unit (ECU), is an electric device that runs a stored computer program that controls the operation of an engine.

ECP
Electrical connection point.

EDA
Emergency driver assistant.

EDS
Explosive destruction system. A system to safely destroy munitions.

EDR
See Event Data Recorder.

Effective Rolling Radius (Re)
For a tire and wheel assembly, the ratio of the linear velocity of the wheel center to the rotational velocity under zero longitudinal slip.

18-Wheeler
A tractor semitrailer with a total of 5 axles and 18 wheels. See Semitrailer; Tractor.

Elastic Deformation
Deformation that is fully recovered after an applied force is removed.

Elastic Impact
An idealized impact where the kinetic energy at separation equals the kinetic energy at the initiation of contact; a fully elastic impact is an impact where the coefficient of restitution is equal to 1.

Electrical Barrier
A device or panel that prevents the passage of a person (or part of a person) or material from one side to another. For more information, see **J2578_AUG2014** [30].

Electromyography (EMG)
The recording of action potentials emitted from contracting muscles. For more information, see *Occupational Biomechanics* [19].

Electronic Control Module (ECM)

See Electronic Control Unit (ECU).

Electronic Control Unit (ECU)

The computer in a vehicle that controls vehicle system operation, including functions such as engine operation, on-board diagnostics (OBD), stability control, safety system operation, among others.

Electronic Data Recorder (EDR)

See Event Data Recorder.

Electronic Stability Control (ESC)

ESC is an extension of an electronic braking system (EBS) that provides protection against loss of control due to excessive vehicle yaw resulting from severe steering maneuvers. The system compensates for oversteer or understeer by comparing the vehicle path intended by the driver to that which is actually occurring. The ESC will then selectively apply braking force to individual wheels of the vehicle or combination vehicle to provide stabilization. For more information, see *Commercial Vehicle Braking Systems: Air Brakes, ABS and Beyond* [5].

EMC

Electromagnetic compatibility. A characteristic of electronic equipment to function acceptably in the intended electromagnetic environment.

Emergency Locking Retractor (ELR)

A retractor incorporating adjustment hardware by means of a locking mechanism that is activated by vehicle acceleration, webbing movement relative to the vehicle, or other automatic action during an emergency and that is capable of withstanding restraint forces when locked. For more information, see **J1803_JAN2013** [16].

EMG

See Electromyography.

EMI

Electromagnetic interference. Unwanted interference or noise in an electrical circuit or path that is caused by an outside source.

End Over (Also Pitch Over)

For motorcycles and bicycles, extreme forward pitching motion, typically resulting in the rider and the vehicle pivoting over the front wheel in the direction of travel and crashing. Also called an endo or a forward flip over. For more information, see **J3133_SEP2019** [18].

End of Event Time

The moment at which the cumulative delta–V within a 20-ms time period changes by 0.8 km/h (0.5 mph) or less, or the moment at which the crash detection algorithm of the airbag control unit resets.

Energy Equivalent Speed (EES)

The speed and corresponding kinetic energy with which a vehicle must contact a fixed rigid object with no rebound for equivalence to conditions of another collision; for example, the energy may be equal to a specified level of residual crush. EES is a preferred term, broader than barrier equivalent velocity (BEV), equivalent barrier speed (EBS), and equivalent test speed (ETS).

Energy Equivalent Velocity

See Energy Equivalent Speed; Equivalent Barrier Speed.

Engine Brake

See Brake; Engine.

Engine Control Module (ECM)

An electronic device in a vehicle (especially heavy trucks) that controls engine operation.

Engine RPM

(1) For vehicles powered by internal combustion engines, the number of revolutions per minute of the main crankshaft of the vehicle's engine; and (2) for vehicles not entirely powered by internal combustion engines, the number of revolutions per minute of the motor shaft at the point at which it enters the vehicle transmission gearbox.

Engine Stop Switch

On a motorcycle, a handlebar-mounted switch that cuts electrical power to the engine ignition system. Sometimes referred to as the engine cutoff switch, emergency stop switch or kill switch (because it "kills" the engine). For more information, see **J3133_SEP2019** [18].

Engine Throttle, Percent Full

The driver-requested acceleration as measured by the throttle position sensor within the engine compared to the fully open position. Note that the throttle position does not necessarily equal accelerator pedal position, for example, the throttle remains open a small amount (e.g., 6%) to allow the engine to operate at idle when the accelerator pedal is at 0% depression.

EPA

Environmental Protection Agency. Among other functions, the EPA works to ensure that Americans have clean air, land, and water.

EPS

Electronic power steering. A system that uses a motor coupled to the steering column or the steering rack to apply a force or torque that aids in the steering of a vehicle. Sometimes referred to as "steer by wire."

ERI

Electronic radiated immunity. A test rating that quantifies the ability of an electronic device to withstand exposure to electromagnetic phenomena.

ESC

See Electronic Stability Control.

ESD

Electrostatic discharge. The sudden flow of electric current between two electrically charged objects.

ESP

An acronym for Chrysler's Electronic Stability Program. See Electronic Stability Control.

Equalizing Hitch (Weight Distributing Hitch)

A mechanical device that connects the trailer to the towing vehicle and by means of leverage applied on both trailer and towing vehicle structures, when properly adjusted, distributes the imposed vertical load at the hitch and coupling connection between structures of towing vehicle and trailer. For more information, see **J684_MAY2014** [40].

Equivalent Barrier Speed (EBS)

The forward speed and corresponding kinetic energy with which a vehicle must contact a flat, fixed, rigid barrier at 90° with no rebound for equivalence to conditions of another collision; for example, the energy may be equal to a specified level of residual crush. See also Energy Equivalent Speed (EES).

Equivalent Test Deformation

See Energy Equivalent Speed.

Equivalent Test Speed (ETS)

ISO term that is not a preferred term. See EBS; EES.

ESC

Electronic stability control.

ESP

Electronic stability program.

ETC

Electronic throttle control. Can also mean electronic toll collection.

ETMS

Enhanced traffic management system.

ETR

Engineering Translation Report. For some General Motors vehicles, the ETR is an additional report to the typical crash data retrieval (CDR) report. Unlike the typical CDR report, the ETR may translate and display data elements for an event data recorder (EDR) system which may or may not be supported on or configured for the vehicle being interrogated.

EU

European Union.

Eutactic Behavior

The intentional use of a driving automation system feature in a manner inconsistent with its design intent, but which is perceived by the user to be safe and based on practical experience. Eutactic behavior is different from forms of abuse such as risk-taking or negligence in that the driver is intentionally acting to balance the risks versus the perceived benefits associated with using the system in a manner that is inconsistent with its design intent. For more information, see **J3114_DEC2016** [32].

EVC

Electronic vehicle control.

Event

A crash or other physical occurrence that causes the trigger threshold to be met or exceeded, or an airbag to be deployed, whichever occurs first.

Event Data Recorder (EDR)

A function within one or more vehicle electronic modules that monitors vehicle and occupant protection system time-series data, prior to and during specific events, defined within the applicable recommended practice, with the intent of retrieving the data after the event. This data is not intended to include freeze-frame-type data (non-EDR data), diagnostic data, or telematics system data. For more information, see **J1698_MAR2023** [41]. Per 49 CFR §563.5 [46]: *Event data recorder (EDR)* means a device or function in a vehicle that records the vehicle's dynamic time-series data during the time period just prior to a crash event (e.g., vehicle speed vs. time) or during a crash event (e.g., delta–V vs. time), intended for retrieval after the crash event. For the purposes of this definition, the event data do not include audio and video data. For more information, see 49 CFR §563.5 [46].

Event Trigger

An event trigger is a set of criteria which, if met, initiates the capture of an event record. For more information, see **J2728_NOV2020** [47].

EVWS

Electric vehicle warning sound. For an electric vehicle, a system designed to alert pedestrians to the presence of vehicles with an electric propulsion system.

EXIF

Exchangeable Image File Format. These files contain digital information about photographic images, such as when a photo was taken, the aperture, shutter speed, and so on.

Extension

Movement of a body appendage that increases the angle between adjoined elements, for example, moving the forearm away from the upper arm at the elbow; the opposite of flexion.

F

FAA

Federal Aviation Administration.

FAF

Freight Analysis Framework. A methodology used by the Bureau of Transportation Statistics, the FAF creates a comprehensive picture of freight movement among states and major metropolitan areas by all modes of transportation.

Fairing

On a motorcycle, a device mounted on the front a motorcycle, whose purpose may include providing the rider some measure of protection from rain, wind, and airborne hazards, reducing drag; modifying appearance; and/or managing airflow to cooling components. Fairings may be attached to either the front caster assembly or frame body assembly, and may incorporate headlamp(s), turn signals, a transparent windshield, accessories, and storage for small items. For more information, see **J3133_SEP2019** [18].

On truck tractors, a fairing is an aerodynamic device mounted to the cab roof to redirect airflow around the sides and over the top of a semitrailer to reduce aerodynamic drag and increase fuel economy.

Fallback-Ready User

The user of a vehicle equipped with an engaged level 3 automated driving system (ADS) feature who is able to operate the vehicle and is receptive to ADS-issued requests to intervene and to evident dynamic driving task (DDT) performance-relevant system failures in the vehicle compelling him or her to perform the DDT fallback. DDT performance by a level 3 ADS assumes that a DDT fallback-ready user is available to perform the DDT as required. There is no such assumption at levels 4 and 5. For more information, see **J3016_JUN2018** [9].

Farm Tractor

A powered farm vehicle designed to pull farm implements (such as a plow, farm trailer, manure spreader, etc.)

FARS

Fatality Analysis Reporting System. FARS is a nationwide census providing NHTSA, Congress, and the American public with annual data regarding fatal injuries suffered in motor vehicle traffic crashes.

FARs

Federal Aviation Regulations.

Fatigue

In materials, the process of accumulating microscopic damage due to cyclic loading that can accumulate with continued cycling until it develops into a crack or other damage that leads to failure of the component. For more information, see *Mechanical Behavior of Materials* [48].

In humans, a state of reduced human physical or mental alertness that impairs performance and is often the result of physical or mental exertion. For more information, see **J3198_202010** [38].

Fatigue Curve

A graph plotting the relationship between cyclic stress amplitude and the number of cycles of a given stress amplitude necessary to produce fracture in a test specimen of a given material; more correctly called a stress-life or S-N curve.

Fatigue Fracture

Fatigue fracture is a progressive type of fracture that may occur under normal service operation in three stages: (1) initiation by a submicroscopic shear or slip mechanism that causes irreversible changes in the crystal structure of the metal; (2) propagation by an increasingly rapid progression of the tip of the fatigue crack in microscopic advances; (3) final rupture, which is final separation or fracture, into two or more parts by a single load application. For more information, see *Understanding How Components Fail* [49].

Fault Tree Analysis (FTA)

FTA is a graphical type of failure analysis used to identify causes of failures of a system or process. FTA starts with what the user experiences and traces back through the system to determine possible alternative causes. See FMEA. For more information, see *Standard Handbook of Machine Design* [50].

FCA

Forward collision avoidance. A system that monitors the speed of a vehicle, the speed of the vehicle in front of it, and the distance between the vehicles to warn the driver of an impending collision.

FCEV

Fuel-cell electric vehicle.

FCW

Forward collision warning. See FCA.

FCWS

Forward collision warning system. See FCA.

FDA

Following distance alert.

Federal Motor Vehicle Standard (FMVSS)

FMVSSs are regulations based on minimum safety performance requirements for motor vehicles or items of motor vehicle equipment. These requirements are specified in such a manner that the public is protected against unreasonable risk of crashes occurring as a result of the design, construction, or performance of motor vehicles, as well as against unreasonable risk of death or injury in the event crashes do occur. For more information, see DOT HS 811 439 [51].

FHWA

Federal Highway Administration. FHWA is a division of the US Department of Transportation that specializes in highway transportation.

Fifth Wheel

The coupling device between a truck tractor and semitrailer; the fifth wheel is typically mounted to the truck tractor (see Figure 21).

FIGURE 21 Rear view of a truck tractor showing the fifth wheel assembly.

vladdon/Shutterstock.com.

First Contact Position

The position, or location, at an accident scene (measured relative to a coordinate system fixed to the earth) of a vehicle, pedestrian, or other object at the time it first has contact with another body in a collision.

First Contact Velocity

The velocity of the center of gravity of a vehicle, pedestrian, or other object at its first contact position.

5-Point Harness

A child restraint harness that has a webbing strap over each shoulder, one on each side of the pelvis, and one between the legs, with all five coming together at a common buckle (see Figure 22). For more information, see **J2939_201304** [2].

FIGURE 22 A 5-point child restraint.

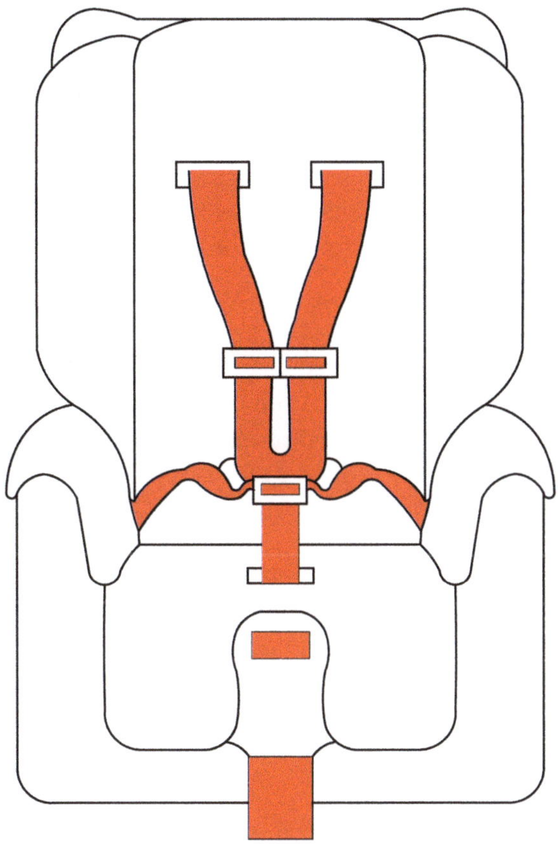

Fixed Object

A stationary object such as a guardrail, bridge railing or abutment, construction barricade, impact attenuator, tree, embedded rock, utility pole, ditch side, steep earth or rock slope, culvert, fence, or building. For more information, see *Manual on Classification of Motor Vehicle Traffic Accidents* [52].

FLC

Forward looking camera.

Flexion

Movement of a body appendage in which the angle between adjoined elements is decreased, for example, moving the forearm toward the upper arm at the elbow; the opposite of extension.

Flip

Movement of a vehicle from a place where the forward velocity of a part of the vehicle suddenly is stopped by an object below its center of gravity, such as a curb, rail, or furrow, with the result that the ensuing rotation lifts the vehicle from the ground. See Vault.

Floor Boards or Floorboards

On a motorcycle, a flat structure(s) mounted to the bottom portion of the motorcycle to provide a secure place for the rider and or passenger to support their feet while riding. Unlike foot pegs that locate the rider's feet in fixed positions, floor boards allow riders to move their feet into different positions for comfort. Foot pegs and floor boards are also known collectively as foot rests. For more information, see **J3133_SEP2019** [18].

FLR

Forward looking radar.

FMCSA

Federal Motor Carrier Safety Administration. An agency in the US Department of Transportation that regulates the nation's trucking industry.

FMCSR

Federal Motor Carrier Safety Regulations. Regulations issued by the Federal Motor Carrier Safety Administration (FMCSA) and compiled in the US Code of Federal Regulations (CFR).

FMEA

Failure modes and effects analysis. An FMEA is a systemized group of activities intended to (1) recognize and evaluate the potential failure of a product/process and its effects, (2) identify actions that could eliminate or reduce the chance of the potential failure occurring, and (3) document the process. It is complementary to the design process of defining positively what the product/process must do to satisfy customers. For more information, see *Potential Failure Mode and Effects Analysis (FMEA) Reference Manual* [53].

FMVSS

See Federal Motor Vehicle Safety Standards. See NHTSA.

Foot Pegs

On a motorcycle, post-like fixtures protruding horizontally from each side (left and right) of the motorcycle intended to support the rider's or passenger's feet by engaging the juncture of the sole of the rider's boot and front vertical surface of the boot heel. The operator's foot pegs are normally located in a position where the rider can operate foot pedals or foot levers without lifting his or her feet from the foot pegs. Foot pegs and floor boards are also known collectively as foot rests. For more information, see **J3133_SEP2019** [18].

Footprint

The portion of the tread of a tire that is in contact with the ground. See Contact Patch. For more information, see *Motor Truck Engineering Handbook* [7].

Fork Assembly

A connected set of components that rotates as a single castering body about the steer axis. It includes the forks, front axle, wheel and tire, front brake, handlebars, front fender, and so on, and in some cases the fairing, lights, and/or windscreen. For more information, see **J3133_SEP2019** [18].

Forward Projection Pedestrian Collision

A frontal collision of vehicle and pedestrian (or possibly a cyclist) where the initial contact area is at or above the height of the center of gravity of the pedestrian (or cyclist) and where a single impact with the frontal geometry of the vehicle causes the pedestrian (or cyclist) to be projected straight forward relative to the vehicle.

Foul (or Foul of a Railroad Track)

Being in proximity of the railroad tracks such that the individual or equipment will be struck by a train or rail service vehicle traversing the track.

Four-Channel ABS

A system that has four sensors and four antilock brake system (ABS) modulator valves (4S/4M). The system may also have six sensors and four ABS modulator valves (6S/4M). For more information, see *Commercial Vehicle Braking Systems: Air Brakes, ABS and Beyond* [5].

Four-Point Transformation

A photogrammetric technique whereby points positioned on a surface reasonably approximated by a plane with unknown locations can be located through the use of four additional points whose locations are known. See Photogrammetry.

4WD

Four-wheel drive.

Foveal Field of View (Central Field of View)

The small, 1- to 2-degree region in the center of the visual field where visual acuity is greatest. For more information, see **J264_OCT2021** [34].

FRA

Federal Railroad Administration. The FRA was created by the Department of Transportation Act of 1966. It is one of ten agencies within the US Department of Transportation concerned with intermodal transportation.

Frame

The backbone structure around which a vehicle is assembled. For more information, see *Motor Truck Engineering Handbook* [7].

Frame Body

Main structural part of a motorcycle that includes, where applicable, the engine and transmission, the fuel tank, a seat and footrests for the rider and a passenger, the battery, and numerous other mechanical components. Attached to the frame body are the front caster assembly, the swing arm, and their related suspension and braking components. For more information, see **J3133_SEP2019** [18].

Free-Rolling Cornering Tire

A loaded-rolling tire moving without applied torque along a curvilinear path at given values of slip angle and/or inclination angle.

Free-Rolling Tire

A loaded rolling tire operated without application of driving or braking torque, that is, longitudinal slip is zero. Under these conditions, the notion that the contact point between the wheel and the roadway is an instant center is valid.

See Free-Rolling Cornering Tire; Straight Free-Rolling Tire. For more information, see *Road Vehicle Dynamics* [8].

Friction

Resistance to sliding over a contact surface between two materials.

Friction Circle (or Friction Ellipse)

A conceptual means to evaluate the limiting behavior of the combined lateral and longitudinal forces developed by a tire and wheel combination operating on a roadway surface with defined friction characteristics.

Friction Coefficient

See Coefficient of Friction.

Frictional Drag Coefficient

An average, uniform (constant) value of a sliding friction coefficient applied to a specific sliding event, such as when an object slides from an initial speed to a stop over a distance, d, or during a speed change, ΔV.

Frictional Drag Factor

See Frictional Drag Coefficient.

Front/Side/Rear Impact Event

For an event data recorder (EDR), a front/side/rear impact event is a crash or other physical occurrence that causes a front/side/rear impact trigger threshold to be met or exceeded or any nonreversible deployable restraint to be deployed, whichever occurs first. For purposes of recording event data, only one front/side/rear impact event can be in progress at any given time. For more information, see **J1698_MAR2023** [41].

Frontal Impact

An impact or collision involving the front of a vehicle.

Frontal Plane

In biomechanics, any plane passing longitudinally through the body from side to side. For more information, see *Basic Biomechanics of the Musculoskeletal System* [3].

FTA

Federal Transit Administration. The FTA provides financial and technical assistance to local public transit systems, including buses, subways, light rail, commuter rail, trolleys, and ferries.

FTA

See Fault Tree Analysis.

Fuel-Cell Module

Fuel-cell modules are comprised of one or more fuel-cell stacks; connections for conducting fuels, oxidants, and exhausts; electrical connections for the power delivered by the stacks; and means for monitoring and/or control. For more information, see **J2578_AUG2014** [30].

Full Brake Application

For a vehicle with air brakes, full brake application means an application of the brake control (treadle or brake pedal) in which pressure in any of the valve's output circuits reaches 85 psi, or the brake control has reached maximum displacement, within 0.2 second after application is initiated. For more information, see **J1626_MAR2023** [54].

Full Throttle

See WOT.

Full Trailer

A towed vehicle with a fixed rear axle and a front axle that pivots that is made to be pulled by a powered tow vehicle (e.g., a farm trailer).

Furrow

A channel in a loose or soft material, such as snow or soil, made by a vehicle tire or some other part of a vehicle moving through the material.

Fusible Link

A device in an electrical path that interrupts current flow when a maximum power throughput occurs. For purposes of high-power electric drive units, a pyrotechnic fusible link may be used to establish a nonreversible and instantaneous break in the current path. For more information, see **J1715_SEP2022** [17].

Fusion

In an automated driving system (ADS), information processing that manages the filtering, correlation, comparison, association, and combination/integration of data of varying uncertainties from multiple sources to produce the highest confidence estimates of the state of the environment to support ADS perception. For more information, see **J3131_202203** [39].

FWD

Front-wheel drive.

G

GA
General aviation. General aviation is defined by the International Civil Aviation Organization as all civil aviation aircraft operations except for commercial air transport or aerial work.

GAW
Gross axle weight. This value represents the total weight carried by an individual axle (front or rear), including the vehicle weight and cargo.

GAWR
Gross axle weight rating. This value represents the maximum allowable weight that can be carried by a single axle (front or rear).

GCW
Gross combination weight. The total weight for a truck or truck tractor, all trailers being towed by the truck or truck tractor, and all equipment and cargo being carried by the truck or truck tractor and trailers.

GCWR
Gross combination weight rating. This value represents the maximum allowable total weight for a truck or truck tractor, all trailers being towed by the truck or truck tractor, and all equipment and cargo being conveyed by the truck or truck tractor and trailers.

Gear Ratio, Axle

Ratio of the speed of the propeller shaft to the speed of the rear axle shaft. For more information, see *Motor Truck Engineering Handbook* [7].

Gear Shift Pedal

On a motorcycle, a foot-operated pedal for selecting transmission gears. The location, operation, and the shift pattern are specified by FMVSS No. 123, Motorcycle Controls and Displays. For more information, see **J3133_SEP2019** [18].

Geometric Stability Ratio (T/(2H))

The value obtained by dividing the average vehicle half-track width (T/2) by the height of the total vehicle center of gravity (H), at a specified load condition. For more information, see **J670_JUN2022** [6].

GFHB

Glare-free high beam.

GHG

Greenhouse gas.

GIF

Graphics Interchange Format. A digital photograph file format.

GIS

Geographic Information Systems. GIS are computer-based tools used to store, visualize, analyze, and interpret geographic data.

Gladhands or Glad Hands

A separable mechanical connector or coupler used to join the service and supply airline hoses when combination vehicles are coupled together. A pair of gladhand connectors resemble a pair of hands shaking when interlocked, hence the name (see Figure 23). For more information, see *Motor Truck Engineering Handbook* [7].

FIGURE 23 A pair of gladhands connected to the front of a semitrailer; in North America, gladhands are color coded. The blue service line supplies air from the brake pedal to the trailer service brakes and the red supply line supplies air to the trailer air tank and releases the trailer's parking brakes.

Thomas Trompeter/Shutterstock.com.

Glare

The light or reflection from a relatively bright light source (compared to the luminance levels in the rest of the visual field). There are two types of glare:

1. Discomfort Glare: Glare of sufficient magnitude so as to cause annoyance or discomfort.
2. Disability Glare: Glare of sufficient magnitude so as to cause a reduction in visual ability.

Depending on the relative intensity of the glare source and the physical condition of the observer,

- Glare can result in annoyance, discomfort, visual fatigue, reduced visual ability, and even temporary "blindness."
- The effect of glare can be magnified by the scattering of light inside the observer's eyes (this problem becomes more prevalent with increasing age).

For more information, see **J264_OCT2021** [34].

Glider or Glider Kit

A glider kit is an incomplete new truck or truck tractor that is delivered without an engine, transmission, or driveline. The name "glider kit" comes from the fact that these units are unpowered from the factory.

GMLAN

A General Motors implementation of the Controller Area Network (CAN)-type serial communication protocol.

Goniometer

Device measuring the angle and range of angular movement between two body segments connected by a joint. For more information, see *Occupational Biomechanics* [19].

Gore

The paved, typically triangularly shaped, area between highway travel lanes and an exit ramp.

Gouge or Gouge Mark

Pavement or ground scar deep enough to be easily felt with the fingers. See Figure 36. See also Chop; Groove.

GPS

Global Positioning System. GPS is a satellite-based infrastructure system that provides users with positioning, navigation, and timing services via a handheld or vehicle-installed receiver.

Grade

The change in elevation per unit distance in a specified direction along the centerline of a roadway or path of travel of a vehicle. It is the difference between the elevations of two points divided by the level distance between the points typically expressed as a percent. For more information, see *Road Vehicle Dynamics* [8].

Groove

A long, narrow pavement gouge or a channel in a pavement.

Gross Combination Weight

The combined weight of a truck tractor and the trailer(s) it is towing, including the weight of all equipment and cargo being conveyed by the tractor and trailer(s).

Gross Combination Weight Rating

The maximum allowable combined weight of a truck tractor and the trailer(s) it is towing, including the weight of all equipment and cargo being conveyed by the tractor and trailer(s).

Gross Vehicle Weight (GVW)

Total weight of a fully equipped truck and payload. For more information, see *Motor Truck Engineering Handbook* [7].

Gross Vehicle Weight Rating (GVWR)

The upper limit of combined weight and cargo for a vehicle established by design, regulation, or both.

Ground Clearance

The distance between the ground and the lowest point of a vehicle between the wheels. For more information, see **J3133_SEP2019** [18].

Guide Loop

In seat belt systems, a load-bearing device through which the seat belt webbing passes and changes direction. It is typically located on the shoulder portion of the seat belt assembly but may be applicable wherever the webbing changes direction and carries restraint loads. Also known as a D-ring, pillar loop, sash guide, or turning loop. For more information, see **J1803_JAN2013** [16].

GVW

See Gross Vehicle Weight.

GVWR

See Gross Vehicle Weight Rating.

H

Handhold

A readily accessible device mounted securely to a vehicle that can be encircled by the fingers of one hand for the purpose of holding on. For more information, see **J2358_AUG2022** [55].

HAPs

Hazardous air pollutants.

Harshness

The higher frequency (30 to 100 Hz) vibrations of the sprung mass. For more information, see **J670_JUN2022** [6].

HDC

Hill descent control. A system on a vehicle that once the system is activated and a speed is set, uses the antilock brake system (ABS) and the traction control system to control wheel slip as a vehicle travels down a slope.

Head Excursion

The distance that the head of a child or crash dummy moves in the direction of impact or on rebound from a crash. For more information, see **J2939_201304** [2].

Head Injury Criteria (HIC)

A parameter defined by NHTSA as

$$\text{HIC} = \left\{ (t_2 - t_1) \left[\frac{1}{t_2 - t_1} \int_{t_1}^{t_2} a(t)\,dt \right]^{2.5} \right\} \max$$

where t_1 and t_2 are the initial and final times (expressed in seconds) of the interval during which HIC attains a maximum value and a(t) is the resultant acceleration (expressed in G) measured at the head CG. For more information, see **J885_FEB2011** [4].

Heading Angle, ψ

The angle between a reference axis fixed in a vehicle and a reference axis fixed in the roadway, giving a measure of vehicle yaw rotation or directional orientation relative to the roadway (see Figure 24).

FIGURE 24 SAE coordinate system (Z-down) showing the vehicle heading angle, ψ, and vehicle sideslip angle, β.

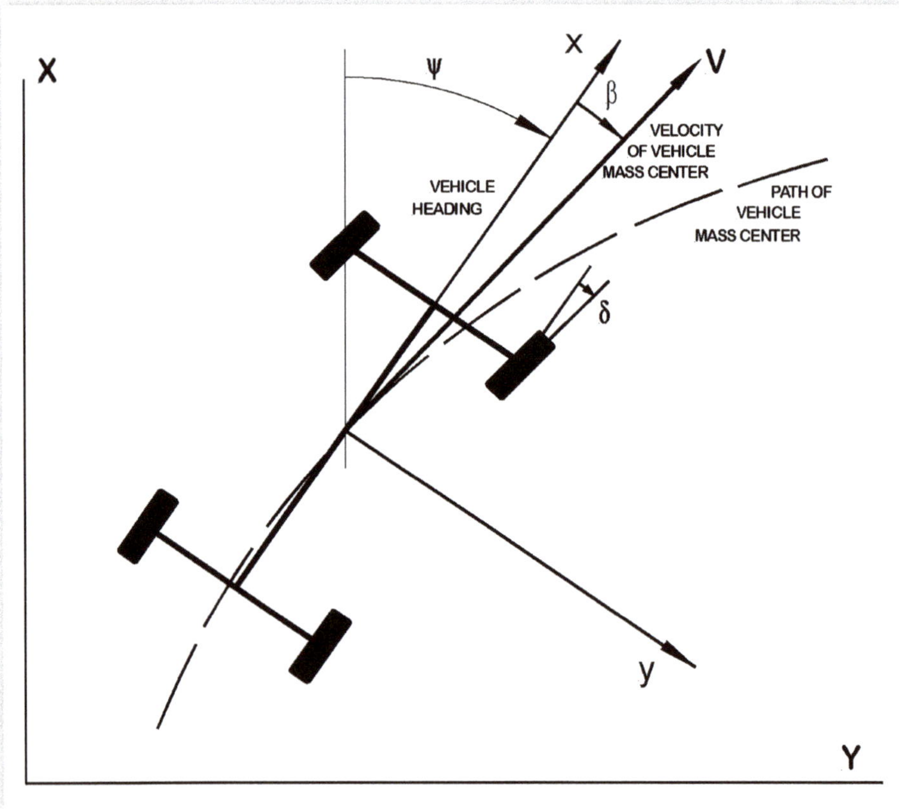

Head-On Impact

Frontal impact where the principal direction of force (PDOF) is at or near zero degrees.

Head-Up Display (HUD)

A virtual display appearing in the area seen through the windshield (see Figure 25). For more information, see **J264_OCT2021** [34].

FIGURE 25 Head-up display.

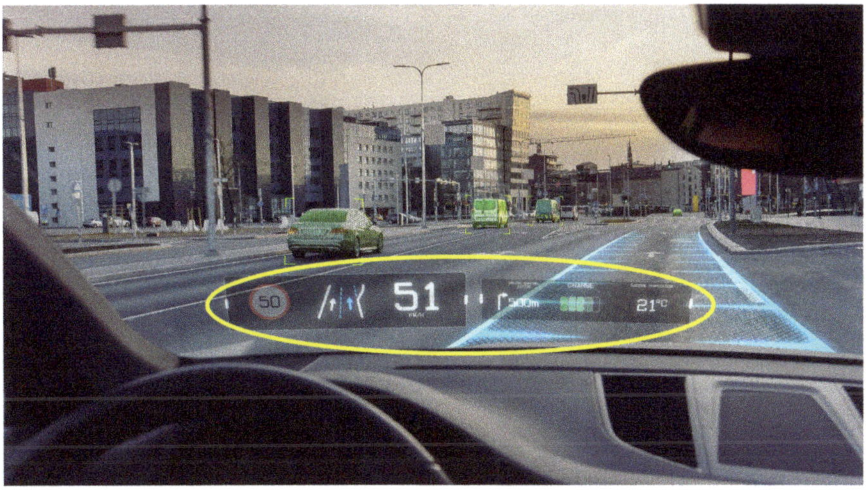

Gorodenkoff/Shutterstock.com.

Heavy Truck (HT)

A motor vehicle designed primarily for transporting cargo with a maximum gross vehicle weight rating (GVWR) over 4,356 kg (10,000 lb). For more information, see **J1100_NOV2009** [14].

Heavy Truck Classifications

See Truck Classifications.

Heavy Vehicle Event Data Recorder (HVEDR)

The HVEDR is an electronic system that captures and records electronic information related to an event during vehicle operation. For more information, see **J2728_NOV2020** [47].

Hematoma

Bruising or bleeding between the brain and skull or elsewhere in the body. For more information, see *Principles of Biomechanics* [25].

HELP
Heavy vehicle electronic license plate. A system that prescreened and prequalified heavy trucks to bypass weight check facilities.

HEV
See Hybrid Electric Vehicle.

High Side
For a motorcycle, a violent vehicle crash mode, where the upper part of the vehicle rapidly rolls upward and toward the direction of travel. For more information, see **J3133_SEP2019** [18].

High Voltage
Voltage levels greater than 30 VAC or 60 VDC. For more information, see **J2578_AUG2014** [30].

Hitch
That part of the connecting mechanism including the ball support platform, the ball, and those components that extend and are attached to the towing vehicle, including bumpers intended to serve as hitches. See Equalizing Hitch; Weight-Carrying Hitch. For more information, see **J684_MAY2014** [40].

HLA
Headlamp assist.

HMIS
Hazardous Materials Information System.

HOS
Heel of shoe. See H-point. For more information, see **J826, J1100_NOV2009** [14].

Hot Shock
Hot shock occurs when a crash force is applied to a vehicle causing the filaments of its headlights, taillights, brake lights, and so on, to become deformed as a result of the filaments being incandescent at the time the force is applied.

HPMS

Highway Performance Monitoring System. The HPMS is a national highway information system that includes data on the extent, condition, performance, use, and operating characteristics of the nation's highways.

H-Point (Hip Point)

Center of rotation between the hip and thigh. Measure of occupant from center hip to top of head restraint is used for seatback height standards for manufacturers. For more information, see *Motor Vehicle Collision Injuries: Biomechanics, Diagnosis, and Management* [20]. Full treatment of design and measurement of the H-point is provided in SAE **J1100**. The devices covered in **J826** provide the means by which passenger compartment dimensions, as defined in **J1100**, can be obtained using a deflected seat rather than a free seat contour as a reference for defining seating space. Figure 26 shows a view of the Hip Point Machine as defined by **J826** annotated with various reference points and characteristics.

FIGURE 26 Diagram of a hip point machine.

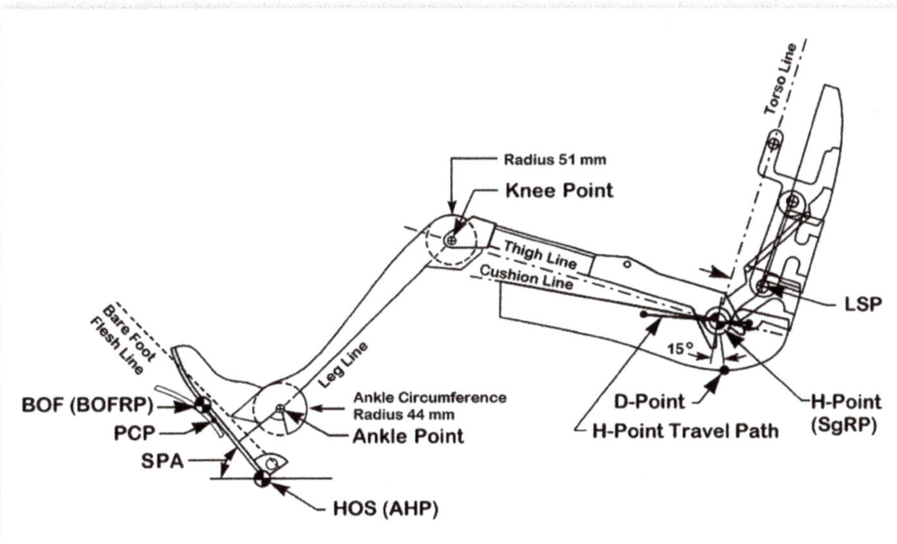

HSI

Human-system integration. HSI is an engineering process and management approach to developing and sustaining systems. It focuses on the comprehensive analysis of interfaces between humans and modern engineering systems.

HSR

High-speed rail.

HTF

Highway Trust Fund. The HTF tracks federal (US) spending and revenue for surface transportation.

Hub Mount Wheel

Wheels that are designed to center on the hub at the bore of the wheel. These wheels generally have straight-through bolt holes, since the bolt holes only supply clearance for the stud. Hub mount wheels are used with two-piece flange nuts.

HUD

See Head-Up Display.

Hurt Report

A report on motorcycle crashes named after the author and principal investigator, Hugh Harrison "Harry" Hurt III. Titled *Motorcycle Accident Cause Factors and Identification of Countermeasures Volume I: Technical Report*, the report was released in 1981 and explored the factors and conditions that led to more than 4,500 motorcycle traffic mishaps in Southern California. For more information, see **J3133_SEP2019** [18].

HV

Heavy vehicle. While definitions and classifications vary across nations, a heavy vehicle is a motor vehicle that has a gross vehicle weight rating (GVWR) greater than 10,000 pounds (44,482 N, 4,536 kg). See Truck Classifications.

Hybrid Electric Vehicle (HEV)

A vehicle that can draw propulsion energy from both of the following sources of stored energy: (1) a consumable fuel and (2) a rechargeable energy storage system (RESS) that is charged by an electric motor-generator system, an external electric energy source, or both. For more information, see **J1715_SEP2022** [17].

Hybrid Vehicle (HV)

A vehicle with two or more energy storage systems, both of which must provide propulsion power, either together or independently. For more information, see **J1715_SEP2022** [17].

Hydraulic Hybrid

A hybrid vehicle where hydraulic fluid is used as the coupling between propulsion elements and energy storage. For more information, see **J1715_SEP2022** [17].

Hydroplaning

A phenomenon whereby a layer of fluid (usually water) on a roadway separates the load-bearing surface of one or more tires of a moving vehicle from the road surface and causes a full loss of traction (longitudinal) and steering (transverse) force components.

Hyperextension

Extreme or excessive extension of a limb or part; backward overbending when applied to the neck. For more information, see **J885_FEB2011** [4].

Hyperflexion

Extreme or excessive flexion of a limb or part; forward overbending when applied to the neck. For more information, see **J885_FEB2011** [4].

I

IBET

Intermodal Bottleneck Evaluation Tool. The IBET provides information and assists transportation planners and policymakers in identifying potential freight bottlenecks in the US transportation system.

ICC

Intelligent cruise control. See Adaptive Cruise Control.

ICE

Internal combustion engine.

Ignition Cycle, Crash

The number (count) of power cycles applied to the recording device at the time when the crash event occurred since the first use of the event data recorder (EDR).

Ignition Cycle, Download

The number (count) of power cycles applied to the recording device at the time when the data was downloaded since the first use of the event data recorder (EDR).

IHBC

Intelligent high beam control. A system that automatically switches the low beam headlamps to high beam mode when no approaching traffic is present.

Illumination

Placement or existence of natural or artificial light on an area presented to a driver.

Imaging

The process by which an event data recorder (EDR) retrieval tool reads an EDR record. For more information, see **J1698_MAR2023** [41].

Impact

(1)The striking of one body against another; (2) short-duration, high-force contact of two objects; (3) a collision of a vehicle with another vehicle, a pedestrian, or some other object. See Collision; Crash.

Impact Center (IC)

The point of intersection of the contact impulse and the intervehicular contact surface for an impact. See Contact Point.

Impact Force (Lever Arm) Moment Arm

See Impulse Moment Arm.

Impact Velocity

The velocity of an object's center of gravity relative to a coordinate system fixed to the earth during an impact. See Postimpact Velocity; Preimpact Velocity.

Impulse

A combination of force, F, and time, τ, defined as a mathematical integral, $\int F d\tau$, of the force over a specific time duration that, when acting on a body, produces a change in the body's momentum.

Impulse Moment Arm

The perpendicular distance from an object's center of gravity to the line of action of an impulse (see **Figure 27**).

FIGURE 27 Illustration of the moment arm, d, of an impulse and the principal direction of force (PDOF).

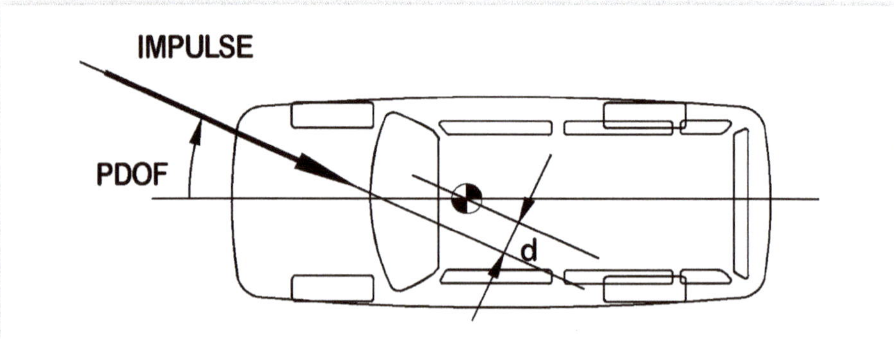

Impulse Ratio

The ratio of the tangential and normal impulse components in planar impact mechanics. See Impulse.

Independent Suspension

A suspension utilizing a separate kinematic control mechanism for the wheel on each side of a vehicle. For more information, see **J670_JUN2022** [6]. See Solid-Axle Suspension.

Independently Controlled Wheel

A directly controlled wheel for which the modulator does not adjust the brake actuating forces at any other wheel on the same axle. For more information, see *Commercial Vehicle Braking Systems: Air Brakes, ABS and Beyond* [5].

Indirectly Controlled Wheel

A wheel at which the degree of rotational wheel slip is not sensed, but at which the modulator of an antilock braking system adjusts its brake actuating forces in response to signals from one or more sensed wheels. For more information, see *Commercial Vehicle Braking Systems: Air Brakes, ABS and Beyond* [5].

Induced Damage

Residual deformation caused without direct contact by virtue of being adjacent to, and physically linked to, the deformation caused by direct contact. See Residual Crush.

Initial Contact

The point in time and space when two objects begin to touch or interact with no significant force; the beginning of an impact.

Injury

Failure or damage of the tissue of a biosystem. For more information, see *Principles of Biomechanics* [25].

INS

Immigration and Naturalization Service.

Instantaneous Center of Motion

The immovable point existing at an instant in time created by one segment (link) of a body rotating about an adjacent segment (without slip); all other points on the body rotate about this immovable point. For more information, see *Occupational Biomechanics* [19].

Instantaneous Driver Workload

The amount of physical and mental activity over a specified unit of time that is required to perform a particular task or set of tasks while driving. For more information, see **J3077_DEC2015** [24]. See Average Driver Workload; Driver Workload; Overall Driver Workload.

Intercooler

Device for cooling air after compression in a turbocharger or supercharger; also referred to as a charge air cooler (CAC). For more information, see *Bosch Automotive Handbook* [56].

Intermodal Transportation

Transportation movement involving more than one mode, for example, rail-motor, motor-air, or rail-water. For more information, see *Motor Truck Engineering Handbook* [7].

Interphalangeal Joints

Joints connecting finger bones. For more information, see *Occupational Biomechanics* [19].

Intervehicular Contact Surface

A single, planar surface that represents the average (over time and space) deformed contact surface between two vehicles or a vehicle and barrier.

Intervehicular Crush Plane

See Intervehicular Contact Surface.

Intrusion

Entry or encroachment of a vehicle's body structure or components of an impacting vehicle into the impacted vehicle's occupant compartment with a reduction of the precrash space.

IPAS

Intelligent parking assist system. A system that assists the driver in parking a vehicle.

IPCC

Intergovernmental Panel on Climate Change. An organization within the United Nations system, based in Geneva, Switzerland.

IRCOBI

International Research Council on the Biomechanics of Impact. Active primarily in Europe, IRCOBI provides an academically rigorous forum for dissemination of the latest research into injury causation and protective systems.

ISA

Intelligent speed adaptation; also intelligent speed advice.

Ischial Tuberosities

Two bony prominences forming the lowest point of the pelvis. On them rests the weight of the body when seated. For more information, see *Occupational Biomechanics* [19].

ISO

International Organization for Standardization. Headquartered in Geneva, Switzerland.

Isotonic

A constant force muscle contraction. For more information, see *Principles of Biomechanics* [25].

ISTEA

Intermodal Surface Transportation Efficiency Act. The ISTEA is a federal law that posed a major change to transportation planning and policy, as the first US legislation on the subject in the post–Interstate Highway System era.

ITS

Intelligent transportation system. An advanced computer-based system intended to make ground transportation safer, more efficient, and less time-consuming in moving people and goods.

J

Jackknife

A loss of control of a tractor trailer in which the tractor's drive wheels lose traction, typically due to excessive braking or acceleration on a slippery surface, and the momentum of the semitrailer pushes forward on the tractor's fifth wheel against rearward longitudinal forces acting farther forward on the tractor. This combination of forces causes the tractor to yaw/rotate one direction or the other about the tractor's front end and the articulation angle between the tractor and semitrailer to decrease rapidly and uncontrollably, often until a rear corner of the tractor's cab contacts the side of the semitrailer.

Jerk

A term used to denote the time rate of change of acceleration of a point. For more information, see *Road Vehicle Dynamics* [8].

Jounce

An upward vertical displacement of the wheel center relative to the sprung mass from a specified reference suspension trim height. For more information, see **J670_JUN2022** [6]. See Rebound.

JPEG

Joint Photographic Experts Group. A digital photograph file format.

K

Kinematics

The branch of mechanics that uses mathematics to describe motion, without reference to the forces or masses involved.

Kinesiology

The study of human movements as a function of the construction of the musculoskeletal system. For more information, see *Occupational Biomechanics* [19].

Kinetic Energy

The energy possessed by a moving body. Kinetic Energy is equal to one half of its mass times the square of its speed.

Kinetics

The branch of mechanics that deals with the motion of a body under the action of given forces.

Kingpin

A round shaft that protrudes downward from the underside of a semitrailer near its forward end that inserts into the truck tractor's fifth wheel. This allows for articulation of the semitrailer with the tractor as the vehicle turns. See Fifth Wheel; Off-Tracking. Also, the round shaft about which the wheel end of a steer axle rotates.

Kingpin Inclination

In a vehicle suspension system, the angle in front elevation between the steering axis and the vertical. For more information, see *Road Vehicle Dynamics* [8].

Kingpin Offset

Kingpin offset at the ground is the horizontal distance in front elevation between the point where the steering axis intersects the ground and the center of tire contact. For more information, see *Road Vehicle Dynamics* [8].

Knuckle (Hub Carrier, Upright)

A suspension part that includes the spindle, to which the control arm(s), link(s), strut, ride spring, and tie rod may be attached. For more information, see **J670_JUN2022** [6].

KPH

Kilometers per hour (also km/h and kph).

KPI

Key performance indicators. A KPI is a critical, quantifiable metric of the status of a process or system relative to an intended result.

L

LA

Lighting automation. A system that switches lights on and off without manual intervention.

LAN

Local Area Network.

Lane Departure Warning (LDW)

Monitors a vehicle's position within the driving lane and alerts the driver as the vehicle approaches or crosses lane markers. This warning is typically multi-modal and can include an auditory alert, visual alert, haptic vibration, or any combination thereof. For more information, see **J3063_MAR2021** [23]. Also LDWS for lane departure warning system.

Lane Keeping Assist (LKA)

Provides steering support to assist the driver in preventing a vehicle from departing the lane. Some systems also help to keep the vehicle centered within the lane. These systems typically provide steering assistance for detected lane markings. For more information, see **J3063_MAR2021** [23].

Lane Splitting

Refers to a motorcycle traveling between traffic lanes with vehicles that are proceeding in the same direction or stopped and at a greater speed than those adjacent vehicles. Lane splitting by motorcycles is illegal in most US states, with one notable exception being California, but is widely used and legal in many other countries. For more information, see **J3133_SEP2019** [18].

LATCH (Lower Anchors and Tethers for Children)

A restraint anchorage system developed to make installation of child restraint systems easier to install.

Latch Plate

A load-bearing device through which the seat belt webbing either passes or is permanently attached. Allows the webbing to change direction and connect with the buckle. Also known as a tongue, tip, or simply as a latch. For more information, see **J1803_JAN2013** [16].

LCA

Lane change assist or lane centering Assist. See Lane Keeping Assist.

LDW

See Lane Departure Warning.

Leading Edge

The foremost part of a vehicle with respect to a vehicle's motion and attitude.

Levels of Driving Automation

The level of driving automation of a vehicle is based on the functionality of the driving automation system, as determined by an allocation of roles in dynamic driving task (DDT) and DDT fallback performance between that system and the (human) user (if any). The manufacturer of a driving automation system determines that system's requirements, operational design domain (ODD), and operating characteristics, including the level of driving automation, as defined below and as shown in Figure 28:

- Level 0: No driving automation
- Level 1: Driver assistance
- Level 2: Partial driving automation
- Level 3: Conditional driving automation
- Level 4: High driving automation
- Level 5: Full driving automation

The manufacturer also defines the proper use of the system. The lower two levels of driving automation (1–2) refer to cases in which the (human) driver continues to perform part of the DDT while the driving automation system is engaged. The upper three levels of driving automation (3–5) refer to cases in which the Automated Driving System (ADS) performs the entire DDT on a sustained basis while it is engaged. For more information, see **J3016_JUN2018** [9].

FIGURE 28 Levels of driving automation.

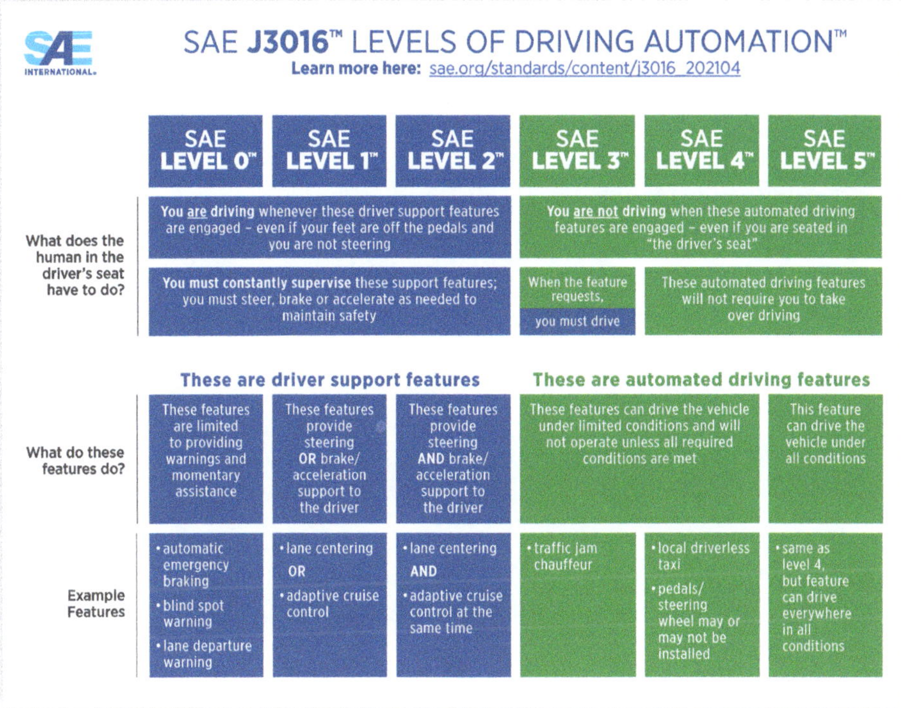

Lidar

An active sensing technology that can measure the distance to, or other properties of, targets (including speed) by illuminating the target with light and analyzing the backscattered light. For more information, see **J3157_FEB2019** [22].

Lift Axle

An unpowered auxiliary axle on a truck, truck tractor, or trailer that can be raised or lowered to change the vertical load distribution of permanent axles. This axle is called a "pusher" if it is positioned ahead of the powered tandem axles, or a tag axle if it is positioned behind the powered tandem axles.

Ligament

Connective tissue attaching bone to bone. For more information, see *Occupational Biomechanics* [19].

Light

Light is radiant energy evaluated with respect to its ability to stimulate the sense of sight in a human observer. In other words, it is visible electromagnetic energy (electromagnetic energy with wavelengths in the range of approximately 400 nm to 700 nm). For more information, see **J264_OCT2021** [34].

Light Truck

An open bed or enclosed motor vehicle designed primarily for transporting cargo with a maximum gross vehicle rating (GVWR) of 4,536 kg (10,000 lb) or less. For more information, see **J1100_NOV2009** [14].

Light Utility Vehicle

Any self-propelled, operator-controlled, off-highway vehicle 1,829 mm (72 inches) or less in overall width, exclusive of added accessories and attachments, operable on three or more wheels or tracks, primarily intended to transport material loads or people, with a gross vehicle weight of 2,500 kg (5,500 pounds) or less, and a maximum design speed less than or equal to 40.23 km/h (25 mph). For more information, see **J2258_JUL2022** [57].

Light Vehicle

An automobile, passenger van, pickup truck, or sport utility vehicle.

Limited-Slip Differential

Mechanical action which resists the free working of an ordinary differential, thus distributing a greater torque to the slower-turning wheel or axle. For more information, see *Motor Truck Engineering Handbook* [7].

LKA

See Lane Keeping Assist.

LKAS

Lane keeping assist system. See Lane Keeping Assist.

LNG

Liquefied natural gas.

Localization

In an automated driving system, the process of estimating the subject vehicle's location, orientation, and associated errors with respect to an external reference frame. For more information, see **J3131_202203** [39].

Locked Event

An event data recorder (EDR) record that is not overwritten by subsequent events. For more information, see **J1698_MAR2023** [41].

Locked Wheel

The wheel traversing a road surface without rotation about the spin axis. For more information, see **J2047_201911** [15].

Locked-Wheel Skid Mark

A skid mark left by a braked, nonrotating wheel sliding (longitudinal slip is zero) in contact with a road surface. For more information, see *Road Vehicle Dynamics* [8].

Locomotion

The act of moving the human body from place to place using the musculoskeletal system. For more information, see *Occupational Biomechanics* [19].

Longitudinal Stiffness

For a vehicle tire and wheel assembly, the absolute value of the first derivative of longitudinal force with respect to longitudinal slip, usually determined at zero longitudinal slip, zero slip angle, and zero inclination angle (zero path curvature). Typically designates C_s. For more information, see **J670_JUN2022** [6] and **J2047_201911** [15].

Lordosis

Excessive curvature of the lumbar or cervical spine.

Low Side

For a motorcycle or bicycle, a crash mode where the upper part of the vehicle rolls down and away from the direction of travel. When a vehicle is leaned over in a turn, the inboard side is closer to the ground. This is the low side of the vehicle. Low-side crashes are usually caused by either locking a wheel due to excessive braking, the front wheel washing out, or the application of excessive power in a turn. It may also be caused by slippery or loose material (such as oil, water, dirt or gravel) on the road surface, especially in a corner. Also called a lay down. For more information, see **J3133_SEP2019** [18].

Low-Speed Vehicle (LSV)

Any powered vehicle with a minimum of four wheels, a maximum level ground speed of more than 32 km/h (20 mph) but not more than 40 km/h (25 mph), and a maximum gross vehicle weight (MGVW) of 1,361 kg (3,000 pounds), that is intended for operating on designated roadways where permitted by law. For more information, see **J2358_AUG2022** [55].

LPG

Liquefied petroleum gas or LP gas.

LSP

Lumbar support prominence. See H-point. For more information, see **J1100_NOV2009** [14].

LTV

Light trucks and vans. See Light Truck; Light Vehicle.

Lugging

Commonly on a motorcycle but on any internal combustion engine powered vehicle, rough or jerking engine operation due to selecting a gear that is too high for road speed. The condition where a motor vehicle's engine speed decreases to the point where the engine does not have enough power to run and drive the vehicle forward smoothly and efficiently. For more information, see **J3133_SEP2019** [18].

Lumbar

Lower back. For more information, see *Principles of Biomechanics* [25].

LV

See Light Vehicle.

M

M2M
Machine to machine. Data communications between electronic devices.

MAIS
See Maximum Abbreviated Injury Scale.

MAIT
Multidisciplinary Accident Investigation Team (NHTSA).

MAP
Manifold absolute pressure.

Master Cylinder
The hydraulic valve assembly used to convert force from the brake pedal to hydraulic pressure for brake actuation. For more information, see **J670_JUN2022** [6].

Maximum Abbreviated Injury Scale (MAIS)
A globally accepted and widely used trauma scale used by medical professionals.

Maximum Crush Depth
Deepest part of a crush profile. See Dynamic Crush; Residual Crush.

Maximum Delta-V, Lateral
The maximum value of the cumulative change in velocity, as recorded by the event data recorder (EDR), of a vehicle along the lateral axis, starting from crash time zero and ending at 0.3 seconds.

Maximum Delta-V, Longitudinal

The maximum value of the cumulative change in velocity, as recorded by the event data recorder (EDR), of a vehicle along the longitudinal axis, starting from crash time zero and ending at 0.3 seconds.

Maximum Delta-V, Resultant

The time-correlated maximum value of the cumulative change in velocity, as recorded by the event data recorder (EDR) or processed during data download, along the vector-added longitudinal and lateral axes.

Maximum Engagement

The point in time when the maximum dynamic crush occurs.

MDB

Moving deformable barrier. Generally used for side impact testing of light vehicles, a moving deformable barrier is a deformable structure, composed of a prescribed honeycomb structure, and mounted to a rolling chassis, that deforms when it collides with a stationary vehicle.

Metacarpal

Refers to the bones of the hand. For more information, see *Principles of Biomechanics* [25].

Metatarsal

Refers to the bones of the feet. For more information, see *Principles of Biomechanics* [25].

MIC

Motorcycle Industry Council. For more information, see **J3133_SEP2019** [18].

Microsleep

Brief, unintentional episodes of lapses in attention that usually occur when a person is fatigued or drowsy. Episodes are often associated with blank stares, periods of prolonged eye closure, and head snapping. Microsleeps can last between a fraction of a second to several minutes and often the person is unaware an event has occurred. Microsleeps often occur when a person's eyes are open, though the person fails to detect and respond to stimuli and events. Driving simulator testing has shown that microsleeps are associated with poor driving performance. For more information, see **J3198_202010** [38].

MID

Message identification (e.g., 136 is the power unit brake system MID). For more information, see *Commercial Vehicle Braking Systems: Air Brakes, ABS and Beyond* [5].

Middle Ordinate

The perpendicular distance between the chord of a circle and the edge of the circle at the center of the chord; this dimension is used when measuring the radius of a curved tire mark when applying the critical speed formula. See Critical Speed Formula.

MIL

Malfunction indicator lamp. A light on the instrument cluster that illuminates if the onboard computer detects a problem with the engine, drivetrain, or electronic system.

Minibike

A very small motorcycle intended for use by a youngster, it typically has only one gear. Originally, these "miniature bikes" were small home-built two-wheelers that used lawn mower engines. Light weight and a low seat height made these little motorized bikes popular with children. These basic bikes are not street legal. For more information, see **J3133_SEP2019** [18].

Minimal Risk Condition

A condition to which a user or an automated driving system (ADS) may bring a vehicle after performing the dynamic driving task (DDT) fallback in order to reduce the risk of a crash when a given trip cannot or should not be completed. For more information, see **J3016_JUN2018** [9].

Minimum Set Speed

Lowest speed at which a cruise control system (adaptive or conventional) can be engaged by the driver. For more information, see **J2399_OCT2021** [10].

MMUCC

Model Minimum Uniform Crash Criteria. Developed by NHTSA, the MMUCC guideline was created to provide state and local agencies with a standard set of motor vehicle traffic crash data variables they should consider collecting during crash investigations.

Modulus of Elasticity (Young's Modulus)

The ratio of stress to strain at any point in the elastic region of a load-deformation curve, yielding a value for stiffness of a material. For more information, see *Basic Biomechanics of the Musculoskeletal System* [3].

Moment of Inertia

A physical property of a body that represents its resistance to rotational acceleration.

Momentum

The momentum of a body is the product of its mass and velocity. Since velocity is a vector quantity, momentum is also a vector quantity.

Monitoring the Driving Environment

The activities and/or automated routines that accomplish real-time roadway environmental object and event detection, recognition, classification, and response preparation (excluding actual response), as needed to operate a vehicle. For more information, see **J3016_JUN2018** [9].

Monitor the User

The activities and/or automated routines designed to assess whether and to what degree the user is performing the role specified for him/her. User monitoring in the context of driving automation is most likely to be deployed as a countermeasure for misuse or abuse (including overreliance due to complacency) of a driving automation system but may also serve other purposes. For more information, see **J3016_JUN2018** [9].

Monitoring Vehicle Performance

For dynamic driving task (DDT) performance-relevant system failures, the activities and/or automated routines that accomplish real-time evaluation of a vehicle's performance, and response preparation, as needed to operate a vehicle. For more information, see **J3016_JUN2018** [9].

Moped

A type of lightweight, low-speed, motorized, two-wheeled, single-track, bank-to-turn vehicle with an auto-stabilizing front steering assembly that generally has a maximum design speed not exceeding 30 mph (50 km/h) and/or an engine displacement of 50 cc or less. Originally, mopeds were equipped with bicycle-like pedals mounted to a crankset that could be used with or instead of motor drive (the source of the term, motor + pedal) to assist the low-power engine when starting or climbing hills. The term "moped" has been increasingly applied by governments to vehicles without pedals but having restricted engine displacement or top speed and/or power output. Although mopeds usually have two wheels, some jurisdictions classify all low-powered three- or four-wheeled motorized cycles as mopeds. For more information, see **J3133_SEP2019** [18].

Motorcoach

A bus that is intended to carry passengers long distances (see Figure 29).

FIGURE 29 A 57-passenger motorcoach.

photo-denver/Shutterstock.com.

Motorcycle

A motorized two-wheeled, single-track, roll-to-turn vehicle with an auto-stabilizing front steering assembly that has a handlebar for steering, a hand-operated throttle control, and a straddle seat for the use of the rider. For more information, see **J3133_SEP2019** [18].

Motor Vehicle

Any vehicle driven or drawn by mechanical power manufactured primarily for use on the public streets, roads, and highways. For more information, see **J2344_OCT2020** [58].

MPG

Miles per gallon.

MPGe

Miles per gallon of gasoline equivalent. This is a measure of the fuel efficiency of vehicles that run on nonliquid fuel (e.g., battery electric vehicles).

MPV

See Multipurpose Passenger Vehicle.

MOD

Moving object detection.

Modulator Valve

Electropneumatic control valve that contains the solenoids used for controlling air pressure. They can precisely modulate air pressure during an ABS-controlled brake application based on signals received from the antilock brake system's electronic control unit (ECU). For more information, see *Commercial Vehicle Braking Systems: Air Brakes, ABS and Beyond* [5].

Motorcycle Category

For motorcycle classification purposes, the general purpose and intended usage for which the motorcycle is designed and equipped. There are five primary categories: on-road, restricted use, dual-purpose, off-road, and competition. For more information, see **J3133_SEP2019** [18].

Motorcycle Crash Modes

There are four primary crash modes for a motorcycle:

1. Motorcycle collides with an object. Most common is another vehicle or a road or roadside hazard.
2. An object collides with the motorcycle. Most common is another vehicle.
3. Motorcycle rotates uncontrollably about the longitudinal axis and the vertical axis. Depending on the precrash lean angle of the motorcycle and other factors, the crash could be a low-side or a high-side crash.
4. Motorcycle rotates (pitches) uncontrollably about lateral axis. Forward rotation is termed an endo. Rearward rotation is called a wheelie.

Motorcycle crashes are complex events that usually involve multiple crash modes before the motorcycle and rider come to a stop. Unlike four-wheeled vehicles, in every motorcycle crash the vehicle and rider will likely impact the ground. For more information, see **J3133_SEP2019** [18].

Motorcyclist

Any person riding on the motorcycle. The motorcyclist in control is referred to as the rider or operator and a motorcyclist riding as a passenger is referred to as a passenger. For more information, see **J3133_SEP2019** [18].

Movement Time

Time interval, usually measured in seconds or milliseconds, for the responding foot or hand to move from one location to another. See Perception-Decision-Reaction Time; Reaction Time. For more information, see **J2944_FEB2023** [28].

MPH

Miles per hour.

MSF

Motorcycle Safety Foundation. The MSF is an American not-for-profit organization, organized in 1972, that creates and promotes education and training for motorcycle riders. It is sponsored by the US manufacturers and distributors of BMW, BRP, Harley-Davidson, Honda, Indian Motorcycle, Kawasaki, KTM, Piaggio/Vespa, Suzuki, Triumph, and Yamaha vehicles.

MTBE

Methyl tertiary-butyl ether. An additive used to maintain the octane ratings of gasoline in the absence of added lead. MTBE is claimed to reduce, through more efficient combustion of the hydrocarbons, the emissions of unburned hydrocarbons during gasoline use. For more information, see *Marks' Standard Handbook for Mechanical Engineers* [59].

MTC

Mechanical throttle control.

Multipiece Rim

A rim consisting of more than one part. Usually two pieces (rim base and side ring) or three pieces (rim base, side ring, and lock ring). For more information, see *Motor Truck Engineering Handbook* [7].

Multipurpose Passenger Vehicle (MPV)

Light truck that has a cargo compartment open to the passenger compartment and is constructed either on a truck chassis or with special features for occasional off-road operation. This category includes sport utility vehicles (SUVs), vans, and crossover vehicles. For more information, see **J1100_NOV2009** [14].

MUTCD

Manual on Uniform Traffic Control Devices. The *Manual on Uniform Traffic Control Devices for Streets and Highways* is issued by the Federal Highway Administration (FHA) of the US Department of Transportation (USDOT) that defines the standards by which traffic control devices, that is, traffic signs and signals and road surface markings, are designed, installed, and used.

MY

Model year. The year used to designate a discrete vehicle model, irrespective of the calendar year in which the vehicle was actually produced, provided that the production period does not exceed 24 months. For more information, see 49 CFR §565.12 Definitions [42].

Myoelectric Signal

Electrical potential produced during contraction of muscle; see Electromyography. For more information, see *Occupational Biomechanics* [19].

N

NASS

National Automotive Sampling System. Overseen by NHTSA, the NASS is composed of two systems: the Crashworthiness Data System (CDS) and the General Estimates System (GES). These are based on cases selected from a sample of police crash reports. CDS data focus on passenger vehicle crashes and are used to investigate injury mechanisms to identify potential improvements in vehicle design. GES data focus on the bigger overall crash picture and are used for problem size assessments and tracking trends.

Navigation

In an automated driving system, the process of locating a vehicle and planning a route from one location to another. For more information, see **J3131_202203** [39].

NCAP

New Car Assessment Program. NCAP is an evaluation program of vehicle crash performance conducted by NHTSA. NHTSA chooses crash test vehicles from passenger car, light truck, sport utility vehicle, and van models that are new, potentially popular (such as the PT Cruiser), redesigned with structural changes, or have improved safety equipment, such as an airbag. The vehicles are tested for frontal collisions, side collisions, and rollover resistance. In addition to providing crash test data, NCAP also provides safety features charts on its Internet website and in its publications that indicate which safety features are found on listed vehicles. For more information, see *The US New Car Assessment Program (NCAP): Past, Present and Future* [60].

NDR

National Driver Register. The NDR maintains the computerized database known as the Problem Driver Pointer System (PDPS) that contains information on individuals whose privilege to operate a motor vehicle has been revoked, suspended, canceled, or denied or who have been convicted of serious traffic-related offenses.

Neutral Steer

Neutral steer occurs when a vehicle traveling on a circular path at constant speed and a constant front-wheel steer angle is accelerated and it remains on a path with the same radius. See Oversteer; Understeer.

Newton's First Law of Motion (Law of Inertia)

An object at rest or in uniform motion will remain at rest or in uniform motion until acted upon by an unbalanced or external force. For more information, see *Road Vehicle Dynamics* [8].

Newton's Second Law of Motion (Law of Acceleration)

The acceleration, a, of a body is directly proportional to the force, F, acting upon the body and is inversely proportional to the mass, m, of the body where $F = ma$. For more information, see *Road Vehicle Dynamics* [8].

Newton's Third Law of Motion (Law of Action and Reaction)

For every action, there is an equal and opposite reaction. Or, whenever one body exerts force on another, the second body always exerts a force on the first body that is equal in magnitude, opposite in direction, and has the same line of action. For more information, see *Road Vehicle Dynamics* [8].

NHTSA

National Highway Traffic Safety Administration. The part of the US Department of Transportation tasked with keeping Americans safe on the nation's roads.

Nondeployment Event

An acceleration event of sufficient magnitude to cause a vehicle's airbag control module (ACM) to record event data but not sufficient to cause the ACM to deploy any airbags or other deployable supplemental restraints, such as seat belt pretensioners.

Nonvolatile Memory

The memory reserved for maintaining recorded event data recorder (EDR) data in a semipermanent fashion. Data recorded in nonvolatile memory is retained after loss of power and can be retrieved with EDR data extraction tools and methods.

NOX (NOx and NO_x)

Nitrogen oxides.

NPTS

Nationwide Personal Transportation Survey. The NPTS serves as the US Department of Transportation's source of data on the amount and nature of personal travel by all modes of transport and the relationship between demographic change and travel.

NTL

National Transportation Library. The NTL was established in 1998 by the Transportation Equity Act for the 21st Century (TEA-21). Administered by the Bureau of Transportation Statistics, the NTL provides US governmental agencies and the public with access to transportation information. The NTL also coordinates creation and dissemination of transportation-related information and provides research and reference services.

NTSB

National Transportation Safety Board. The NTSB is a federal agency charged with investigating every civil aviation accident in the United States and significant events in other modes of transportation—railroad, transit, highway, marine, pipeline, and commercial space.

O

OAL

Overall length. For a light vehicle, the length of the vehicle from the front of the front bumper to the rear of the rear bumper; for a truck, the distance from the front bumper to the end of the frame. See Figure 3.

OAW

Overall width of a vehicle. The value is frequently specified as to whether or not it includes the side-view mirrors.

OBD

On-board diagnostics. Specification of diagnostic services and functionally addressed request/response messages required to be supported by motor vehicles and external test equipment for diagnostic purposes that pertain to motor vehicle emission-related data. These functions of a vehicle are controlled by SAE Standards **J1978** and **J1979**.

OBD-II

On-Board Diagnostics, version 2. See OBD.

Object and Event Detection and Response

The subtasks of the dynamic driving task (DDT) that include monitoring the driving environment (detecting, recognizing, and classifying objects and events and preparing to respond as needed) and executing an appropriate response to such objects and events (i.e., as needed to complete the DDT and/or DDT fallback). For more information, see **J3016_JUN2018** [9].

Oblique Impact

An impact in which the contact impulse does not pass through the center of gravity. See Central Impact.

Occipital

Refers to the back of the head. For more information, see *Principles of Biomechanics* [25].

Occupant Compartment

That portion of a vehicle's interior designed for the use of passengers during operation of the vehicle.

Occupant Size Classification

For the right front passenger, the classification of the occupant as an adult and not as a child, and for the driver, the classification of the driver as not being of small stature.

ODD

See Operational Design Domain.

ODI

Office of Defects Investigation. The ODI administers the NHTSA's safety recall program.

OEDR

See Object and Event Detection and Response.

OEDR Reaction Time

The time it takes for the automated driving system (ADS) to initiate a measurable response following the onset of an initiating event in the context of scenario-based testing in a controlled environment (e.g., track testing or simulation). For more information, see **AVSC00006202103** [61].

OEM

Original equipment manufacturer. Normally used in reference to the manufacturer of the total vehicle rather than the builder of a component. For more information, see *Commercial Vehicle Braking Systems: Air Brakes, ABS and Beyond* [5].

Offset

The distance between the longitudinal axes of two vehicles in frontal contact (see Figure 30). See Overlap.

FIGURE 30 Illustration of offset and overlap in a vehicle collision.

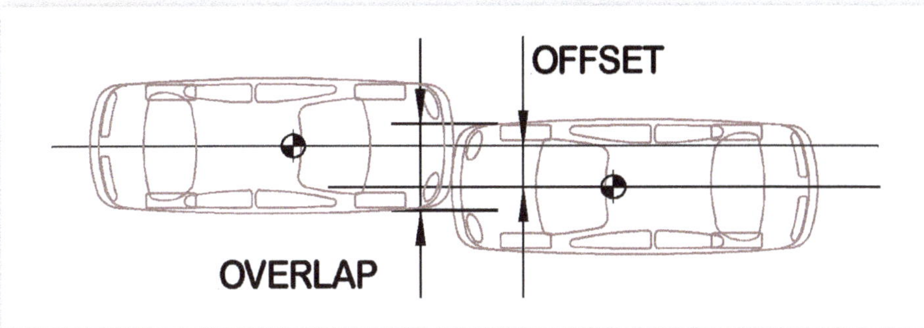

Off-Tracking

A dynamic of articulated vehicles that occurs in a turning tractor-trailer combination vehicle whereby the trailer tires track along a different radius than the turn center of the towing vehicle tires. The trailer tires usually track a smaller radius at lower speeds and a larger radius at high speeds. The extent of off-tracking is also affected by the overall length of the combination vehicle, with longer vehicles off-tracking more than shorter ones at a given speed (see Figure 31).

FIGURE 31 Off-tracking of a tractor and semitrailer (low speed).

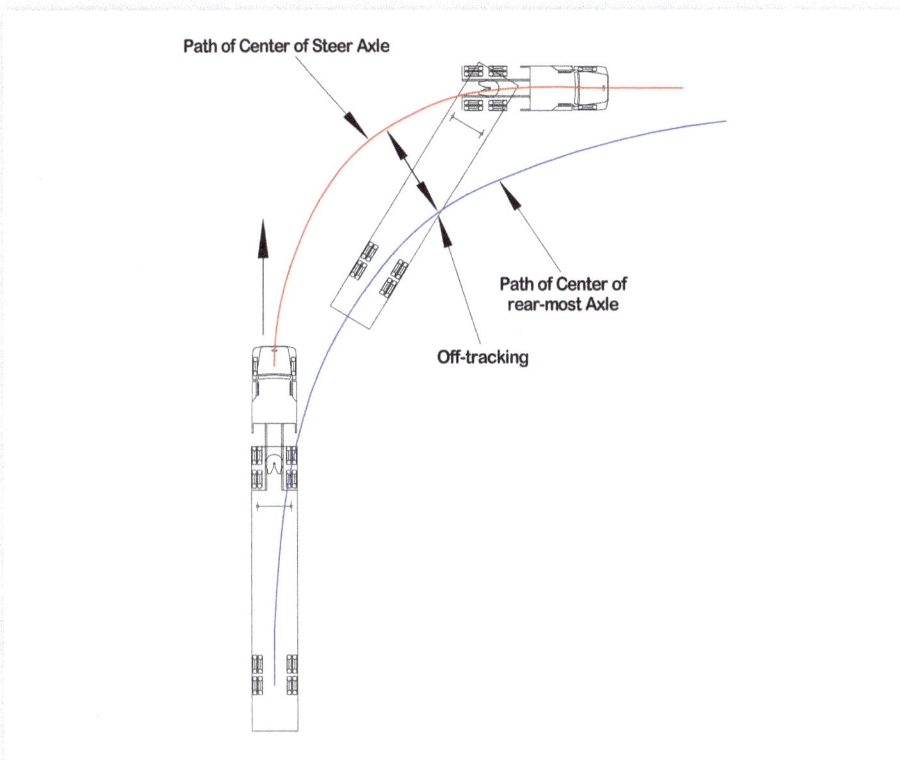

On-Road Rollover

A rollover that initiates on a road surface. For more information, see **J2926_201707** [36].

OPEC

Organization of Petroleum Exporting Countries.

Operate (a Motor Vehicle)

Collectively, the activities performed by a (human) driver (with or without support from one or more level 1 or 2 driving automation features) or by an ADS (levels 3–5) to perform the entire dynamic driving task (DDT) for a given vehicle during a trip. Sometimes the word "drive" is used in lieu of "operate" as it pertains to the actions of the operator of a vehicle. For more information, see **J3016_JUN2018** [9].

Operational Design Domain (ODD)

Operating conditions under which a given driving automation system or feature thereof is specifically designed to function, including, but not limited to, environmental, geographical, and time-of-day restrictions, and/or the requisite presence or absence of certain traffic or roadway characteristics. For more information, see **J3016_JUN2018** [9].

Orbital

Refers to the eye socket. For more information, see *Principles of Biomechanics* [25].

ORC

Occupant restraint controller. See ACM.

OTR

Over the road.

Overall Driver Workload

The amount of physical and mental activity that is required to perform a particular task or set of tasks while driving. For more information, see **J3077_DEC2015** [24]. See Average Driver Workload; Driver Workload; Instantaneous Driver Workload.

Overhang, Front/Rear

The longitudinal dimension of a vehicle from the center of the front/rear wheels to the foremost/rearmost point on the vehicle, including bumper, bump guards, tow hooks, and/or rub strips if standard equipment.

Overlap

The length of mutual contact damage. See Figure 30. See Offset.

Overreliance

In an automated driving system, the situation in which, despite apparently degraded driving automation system performance, the user does not intervene because the system has not issued a warning or otherwise seems to be performing adequately. For more information, see **J3114_DEC2016** [32].

Override

A condition in a collision where the main structural members, such as the bumper or chassis frame of the striking vehicle, are above the main structural members, such as the bumper or chassis frame of the struck vehicle (see Figure 32). See Underride.

FIGURE 32 Illustration of override in a vehicle-to-vehicle collision.

Oversteer

When a vehicle traveling on a circular path at constant speed and constant front wheel steer angle is accelerated, and tends to decrease its radius, that is defined as oversteer, respectively. See Neutral Steer; Understeer.

Overturning Moment

The moment exerted on the tire by the road about the X_T-axis. See Figure 4. For more information, see **J2047_NOV2019** [15].

P

PAEB
Pedestrian automatic emergency braking. A vehicle system that aids the driver in avoiding collisions with pedestrians.

PAR
Police accident report.

Parking Brake
A device to prevent the movement of a stationary vehicle. See Service Brake System. For more information, see **J2258_JUL2022** [57].

Parking Collision Warning (PCW)
Detects objects close to a vehicle during parking maneuvers and notifies the driver. For more information, see **J3063_MAR2021** [23].

Passenger
An occupant in a vehicle who has no role in the operation of that vehicle. The person seated in the driver's seat of a vehicle equipped with a level 4 automated driving system (ADS) feature designed to automate high-speed vehicle operation on controlled-access freeways is a passenger while this level 4 feature is engaged. This same person, however, is a driver before engaging this level 4 ADS feature and again after disengaging the feature in order to exit the controlled access freeway. For more information, see **J3016_JUN2018** [9].

Passenger Car
A motor vehicle designed for carrying ten or fewer persons, excluding multipurpose vehicles, motorcycles, and trailers. For more information, see **J1100_NOV2009** [14].

Pathology

The discipline dealing with the development and description of disease in terms of altered structure and function of the body. For more information, see *Occupational Biomechanics* [19].

Payload

Actual weight of useful cargo carried by vehicle. For more information, see *Motor truck Engineering Handbook* [7].

Payout

The length of seat belt webbing extracted out of a retractor during loading. It may consist of webbing distance to lock, film spooling, webbing stretch, and load limiting. For more information, see **J1803_JAN2013** [16].

PCM

Powertrain control module; also pulse code modulation (a way of digitally transmitting analog data). See Engine Control Module.

PCR

Police crash report.

PCW

See Parking Collision Warning.

PD

Pedestrian detection; also PDS—Pedestrian detection system.

PDOF

See Principal Direction of Force.

PDR Time

See Perception-Decision-Reaction Time.

Peak Friction Coefficient

The ratio of the maximum value of longitudinal force to the simultaneous vertical force occurring prior to wheel lockup, as the braking torque is progressively increased, as measured using the procedure in ASTM E1337. For more information, see **J1626_MAR2023** [54].

The maximum values of the tire shear force coefficient attainable by a tire with a given set of steady-state operating conditions: (1) during the application of a braking torque at zero slip angle; (2) during the application of a driving torque at zero slip angle; or (3) when cornering with zero applied wheel torque. Typically, the peak coefficients of friction determined in the three pure modes of braking, driving, and cornering will have different values; moreover, the values for cornering may differ for positive and negative slip angles. For more information, see **J670_JUN2022** [6].

Pedestrian Center of Gravity (Pedestrian CG)

To find the approximate location of the center of gravity for an adult pedestrian, multiply the person's height (without shoes) by 0.57 and then add the shoe heel height. For more information, see *Road Vehicle Dynamics* [8].

Perception

The process of detecting some object or situation and comprehending its significance. For more information, see *Road Vehicle Dynamics* [8]. In an automated driving system (ADS), the capability of the system to sense and characterize the entities, events, and situations in its environment. For more information, see **J3131_202203** [39].

Perception-Decision-Reaction Time (Response Time)

The time required by a person to complete a response to an event or stimulus; also the sum of reaction time and movement time. See Movement Time; Reaction Time.

Perception-Response Time (PRT)

Time interval, usually measured in seconds or milliseconds, between when an initiating event (e.g., obstacle in the roadway) can be physically sensed and when braking begins. This term is often called perception-reaction time in roadway design and accident reconstruction literature, even though it involves an intermediate movement to the brake and is more than just reaction time as defined here. PRT includes the time needed to (1) see/perceive the roadway element or obstacle, (2) to complete relevant cognitive operations (e.g., recognize hazard, read sign, decide how to respond, etc.), and (3) initiate a maneuver (e.g., take foot off accelerator and step on brake pedal). For more information, see **J2944_FEB2023** [28]. See Perception-Decision-Reaction Time.

Peripheral Field of View

The outer, non-foveal field of view. For practical purposes, anything beyond the small (1 to 2 degrees) central area of vision can be considered part of the peripheral field of view. This area is characterized by (a) poor acuity, (b) poor color discrimination, and (c) optimal sensitivity to low levels of light. For more information, see **J264_OCT2021** [34]. See Foveal Field of View.

PEV

See Plug-in Hybrid Electric Vehicle.

PFD

Personal Flotation Device. A jacket that provides buoyancy while a person is in the water.

PHEV

See Plug-in Hybrid Electric Vehicle.

Photogrammetry

The process of determining the quantitative dimensional information of objects in two or three dimensions through the process of recording, interpreting, and transforming measurements from a flat photographic image.

Physiological Response

The normal response and adaptation of the living organ and its parts to stress. For more information, see *Occupational Biomechanics* [19].

Physiology

The science that deals with the normal function of the living organ and its parts. For more information, see *Occupational Biomechanics* [19].

PID

Parameter identification (e.g., 194 is the Transmitter System Diagnostic Code and Occurrence Count PID). For more information, see SAE **J1587/J1708** and *Commercial Vehicle Braking Systems: Air Brakes, ABS and Beyond* [5].

PIM

See Planar Impact Mechanics.

Pintle Hook

Coupling at rear of truck for the purpose of towing trailer or other units. For more information, see *Motor Truck Engineering Handbook* [7].

Pitch, Roll, Yaw

Terms that distinguish rotations of a vehicle about three perpendicular axes with origin at the vehicle's center of gravity. Pitch is rotation about the horizontal, side-to-side axis; roll is rotation about a horizontal front-to-rear axis; and yaw is rotation about the vertical axis. See Figure 42. See Yaw Angle.

Pitman Arm

Steering gear arm attached to drag link. For more information, see *Motor Truck Engineering Handbook* [7].

Planar Impact

An impact in which all forces, moments, and motion of the vehicle(s) take place in a plane.

Planar Impact Mechanics (PIM)

Analysis of a planar impact using a model in which the vehicles are considered to be rigid bodies, that is, they have rotational inertia, meaning the resulting model includes rotational effects.

Plantar Flexion

Bending about the ankle joint in the direction of the sole of the foot. For more information, see *Occupational Biomechanics* [19].

Plastic Impact

An impact with little or no rebound at the end of impact. A perfectly plastic impact is where the coefficient of restitution is equal to zero.

Plow

The limit condition for vehicle directional response wherein the front tires have reached their cornering limit while the rear tires have not reached their limit. For more information, see **J670_JUN2022** [6]. See Drift; Spin.

Plug-in Hybrid Electric Vehicle (PHEV)

An electric vehicle that can be recharged with an off-board source of electricity; it includes both battery electric vehicles (BEVs) and plug-in hybrid electric vehicles (PHEVs). For more information, see **J1715_SEP2022** [17].

Plysteer

Plysteer is a lateral force and aligning moment that cause a tire to behave as if it had an inbuilt steer angle. The individual tire acts as if it always steers slightly in a particular direction. The cause of this behavior is structural anisotropy due the set of cord angles, ply layups, and tread sculpture elements, which exists within a given tire design. Plysteer is primarily a design effect. For more information, see **J2047_201911** [15].

PM-10

Particulate matter of 10 microns in diameter or smaller.

PM-2.5

Particulate matter of 2.5 microns in diameter or smaller.

PMT

Passenger-miles of travel. Generally used in aviation, the sum total of the distances traveled by each passenger for the leg of a flight.

Pneumatic Timing

The time required for transmission of air to or from brake chambers upon brake actuation or release. For more information, see *Motor truck Engineering Handbook* [7].

Pneumatic Tire

A flexible, hollow semi-toroid mounted on the rim, filled with compressed gas (usually air); it becomes load-bearing when filled with compressed gas (usually air). For more information, see **J670_JUN2022** [6] and **J2047_201911** [15].

PNG

Portable Network Graphics. A digital photograph file format.

Pocket Bike

These are minibikes that are designed to look like sport bikes. The usual height is less than 20 inches (50 cm), with a length up to 3 feet 3 inches (1 m). Maximum speed varies between 19 to 40 mph (30 to 64 km/h). Pocket bikes are typically equipped with either small displacement engines, with a maximum of 4.5 to 6 horsepower (3.4 to 4.5 kW), or battery-powered electric motors. For more information, see **J3133_SEP2019** [18].

Point of Contact

The point of intersection of the contact impulse and the intervehicular contact surface during an impact. See also First Contact Position; Impact Center; Principal Direction of Force.

Point of Impact (POI)

(1) The collision sequence where involved objects first come into one another or a surface; can be referred to as the area of impact (AOI) and the initial contact. (2) The location on the road or other surface where objects such as vehicles collided. See also Area of Impact (AOI). For more information, see *Road Vehicle Dynamics* [8].

Point Mass

An idealized concept from mechanics whereby an object is considered to have mass but no extent and no finite dimensions, and as a consequence, its rotation is irrelevant. See Rigid Body.

Point Mass Impact Mechanics

Analysis of a collision in which the vehicles are considered point masses, that is, they have no rotational inertias. This impact model neglects the rotational motion of the vehicles.

Point of Rest

Controlled or uncontrolled postcollision position and orientation of a vehicle or person. For more information, see *Road Vehicle Dynamics* [8].

Pole Trailer

A pole trailer is a trailer that is attached to a towing vehicle by a telescoping or sliding pole.

Postcollision Trajectory

See Postimpact Trajectory.

Postcrash Damage

Damage to a vehicle after it came to rest, including damage that results during rescue, towing, and salvage operations.

Postimpact Speed

The magnitude of the velocity of an object in a collision at the time of separation, or end of contact. See Postimpact Velocity; Separation Speed.

Postimpact Trajectory

The path of the center of gravity of a vehicle from separation to rest. This trajectory is typically defined in an earth-fixed coordinate system. For more information, see *Road Vehicle Dynamics* [8].

Postimpact Velocity

The velocity of an object in a collision at the time of separation or end of contact. See Postimpact Speed; Separation Speed.

Power Factor

The ratio of active (or real) power in watts to apparent power in volt-amps at the fundamental frequency of power transmission (i.e., 60 Hz). It is a measure of the phase shift that occurs between line voltage and line current when the AC line is loaded with a linear load having reactive characteristics. For more information, see **J1715_SEP2022** [17].

Power Takeoff

Means to transmit power from engine to auxiliary equipment rather than the drivetrain.

Powertrain

Engine plus drivetrain. For more information, see *Motor Truck Engineering Handbook* [7].

Preimpact Velocity

The velocity of a vehicle in a collision at the instant of its initial contact.

Pretensioner

A device or mechanism incorporated into a seat belt system that deploys under predetermined crash criteria and is designed to tighten the webbing. Typical application locations include the retractor, buckle, or outboard lap anchor. Some systems may incorporate multiple pretensioners in any combination. For more information, see **J1803_JAN2013** [16].

Primary Line of Sight

The line connecting the point of observation and the fixation point. (The point of observation is the midpoint of a line connecting the centers of rotation of the two eyes.) For more information, see **J264_OCT2021** [34].

Principal Direction of Force (PDOF)

The direction of the line of action of the contact impulse in a planar collision expressed in degrees, measured clockwise from the forward, longitudinal axis (heading) of a vehicle. See Figure 27.

Prone

Lying face down on the stomach. For more information, see *Principles of Biomechanics* [25].

Propeller Shaft

Driveshaft used to transmit torque from the engine or transmission to the rear axle. For more information, see *Motor Truck Engineering Handbook* [7].

Proportioning Valve

A hydraulic valve that reduces brake pressure supplied to the rear brakes, relative to front brake pressure, to maintain brake balance in the presence of longitudinal load transfer. For more information, see **J670_JUN2022** [6].

PRNDL

Park-Reverse-Neutral-Drive-Low. The conventional shift mechanism sequence for a vehicle with an automatic transmission.

PSMD

Parking slot marking detection. A vehicle-based system that detects markers on a roadway or parking lot surface to identify the position of a parking slot.

PTC

Positive train control. PTC systems are designed to prevent train-to-train collisions, overspeed derailments, incursions into established work zones, and movements of trains through switches left in the wrong position.

PTO

See Power Takeoff.

Pup or Pup Trailer

A short semitrailer, typically used in combination with a converter dolly and another semitrailer to create a double trailer combination or two converter dollies and two trailers to create a triple trailer combination.

Pusher Axle

A non-driven (dead) axle installed forward of the driven axle(s) to increase the permissible gross weight, and consequently the payload. For more information, see *Motor Truck Engineering Handbook* [7]. See Drop Axle; Lift Axle; Tag Axle.

PUV

Personal-use vehicle.

Q

Quick-Release Valve

In vehicles with pneumatic brake systems, the function of the quick-release valve is to speed up the exhaust of air from the air chambers. It is mounted close to the chambers it serves. For more information, see *Air Brake Handbook* [62].

R

Rack and Pinion Steering Gear

A steering gear employing a rack and pinion gear set. This design transfers rotational motion of the input shaft to translational motion of the steering rack. It is used with laterally orientated tie rods connected directly to the rack. For more information, see **J670_JUN2022** [6].

Radius of Gyration

The square root of the quotient of the moment of inertia and the mass of a rigid body. See Moment of Inertia.

Rake Angle

On a motorcycle, generally the angle between the steer axis and vertical vehicle axis when the vehicle is upright with the steering centered. Contributes to directional stability. Similarly, for vehicle windshields, rake angle is the angle between the transverse parallel plane of the windshield and the vehicle's vertical axis; also called "windshield rake" (see Figure 33). For more information, see **J3133_SEP2019** [18].

FIGURE 33 Rake angle of a motorcycle.

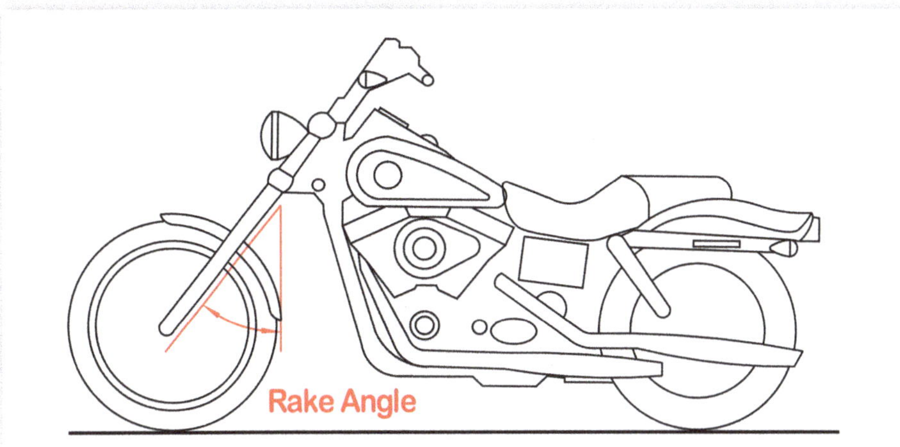

Range of motion

The range of translation and rotation of a joint for each of its six degrees of freedom. For more information, see *Basic Biomechanics of the Musculoskeletal System* [3].

Range

The maximum distance that a vehicle can travel without additional energy in the form of fuel or off-vehicle charging. For more information, see **J1715_SEP2022** [17].

RB

See Regenerative Braking.

RCM

Restraint control module. Nomenclature used by the Ford Motor Company for an airbag control module. See ACM.

RCTA

Rear cross-traffic alert. See CTA.

Reaction Time

Time interval, usually measured in seconds or milliseconds, from onset of an initiating event to the first observable response to that event, such as a movement of the driver's hand (on the steering wheel) or foot (on a pedal or from the floor); the beginning or end of an utterance by the driver for voice-activated controls. For more information, see **J2944_FEB2023** [28]. See Movement Time; Perception-Decision-Reaction Time.

Rebound

A downward vertical displacement of the wheel center relative to the sprung mass from a specified reference suspension trim height. See Jounce. For more information, see **J670_JUN2022** [6].

Rechargeable Energy Storage System (RESS)

Any energy storage system that has the capability to be charged and discharged. Examples include batteries, capacitors, and electromechanical flywheels. For more information, see **J1715_SEP2022** [17].

Recirculating-Ball Steering Gear

A steering gear generally employing a ball screw mechanism to transfer rotational motion from the input shaft to rotational motion of the output shaft. For more information, see **J670_JUN2022** [6].

Reconstruction

A systematic process of evaluating evidence associated with a collision sequence and applying accepted physical principles to determine how the collision occurred. For more information, see *Road Vehicle Dynamics* [8].

Reefer

Semitrailer with a heavily insulated refrigerated body and refrigeration unit. Also used colloquially to refer to the combination of a truck tractor with a refrigerated semitrailer. For more information, see *Motor Truck Engineering Handbook* [7].

Regenerative Braking (RB)

The function of the electrical machine in converting mechanical power inputs to electrical power outputs, ultimately for the purpose of providing energy for storage in the rechargeable energy storage system (RESS) and to provide assisted braking force. For more information, see **J1715_SEP2022** [17].

Reinforced Insulation

A single insulation system with such mechanical and electrical qualities that it, in itself, provides the same degree of protection against the risk of electric shock as does double insulation. The term "single insulation system" does not necessitate that the insulation must be in one homogeneous piece. The insulation system may comprise two or more layers that cannot be tested as supplementary or basic insulation. For more information, see **J2578_AUG2014** [30].

Release Timing

The measurement in time from initial brake movement to reach 5 psi from 95 psi pressure in the brake chambers. For more information, see *Commercial Vehicle Braking Systems: Air Brakes, ABS and Beyond* [5].

Remote Driver

A driver who is not seated in a position to manually exercise in-vehicle braking, accelerating, steering, and transmission gear selection input devices (if any) but is able to operate the vehicle. For more information, see **J3016_JUN2018** [9].

Remote Parking Assistance

Without the driver being physically present inside the vehicle, this provides steering, braking, accelerating, and/or gear selection while moving a vehicle into or out of a parking space. The driver must constantly supervise this support feature and maintain responsibility for parking. For more information, see **J3063_MAR2021** [23].

Request to Intervene

Notification by an automated driving system (ADS) to a fallback-ready user indicating that she or he should promptly perform the dynamic driving task (DDT) fallback, which may entail resuming manual operation of the vehicle (i.e., becoming a driver again) or achieving a minimal risk condition if the vehicle is not drivable. For more information, see **J3016_JUN2018** [9].

Residual Crush

The permanent deformation formed by the nominal external surface of a vehicle caused by an impact, usually measured relative to the corresponding as-manufactured undeformed surface. See Crush Area; Crush Profile.

RESS

See Rechargeable Energy Storage System.

Rest Position

The location of the center of gravity of a vehicle following an accident measured relative to a coordinate system fixed in the earth.

Retreaded Tire

A tire to which a new tread has been affixed to an existing carcass or casing in place of the initial tread. For more information, see **J2047_201911** [15].

Reverse Projection Photogrammetry

The photogrammetric procedure of inserting a transparency that contains outlines of transient and fixed objects into a camera for the purpose of determining the position and orientation of the camera at the time the original photograph was taken to facilitate the relocation of the transient information.

RFG

Reformulated gasoline.

RFI

Radio frequency interference. See EMI.

RICSAC

Research Input for Computer Simulation of Automobile Collisions. The acronym given to a series of vehicle-to-vehicle tests sponsored by the US Department of Transportation and conducted by the Calspan Corporation in the mid- to late 1970s. These tests were the first collisions with comprehensive data collection. The data from some of these crashes were used for a variety of purposes including the validation of many collision reconstruction models and some vehicle postimpact trajectory models. The reports for these collisions, in four volumes, are available on the internet.

Rigid Body

A concept from mechanics whereby an object is considered to have mass and dimensions (such as length, width, radius, etc.) that remain constant and which provide resistance to rotation. See Moment of Inertia; Point Mass.

Rim

That part of the wheel on which the tire is mounted and supported. See item 9 in Figure 19. For more information, see **J2047_201911** [15].

Rim Flange

That part of the rim which provides lateral support to the tire (see Figure 34). For more information, see **J2047_201911** [15].

FIGURE 34 Rim flange and its various characteristics and parameters.

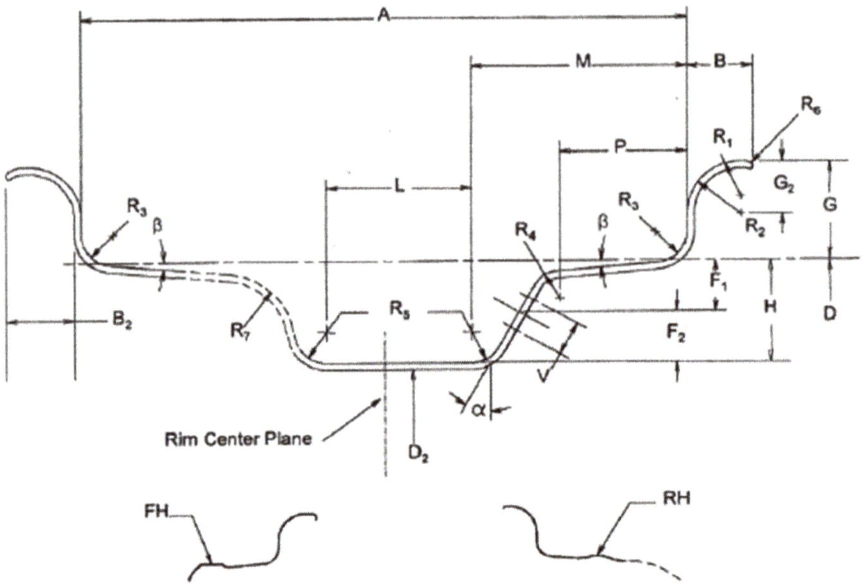

A Specified rim width
B Flange width
B_2 Flange offset
D Specified rim diameter
D_2 Rim inside diameter
F_1, F_2 Valve hole location
FH Flat Hump
G Flange height
G_2 Flange radius location
H Well depth
L Well width
M Well position
P Bead seat width
R_1 Flange compound radius
R_2 Flange radius
R_3 Bead seat radius
R_4 Well top radius
R_5 Well bottom radius
R_6 Flange edge radius
R_7 Well wall radius
RH Round Hump
V Valve hole
α Well angle
β Bead seat angle

Rim Size Designation

Rim diameter designation X rim contour designation. Example 17 X 6.5 J, which denotes a 17-inch nominal rim diameter, 6.5-inch nominal rim width, and J rim profile. For more information, see **J2047_201911** [15].

Road Operator

A traffic participant who provides, operates, and maintains the roadways and supporting infrastructure that enable and support the mobility needs of road users. For more information, see **J3216_JUL2021** [11].

Road Plane

A plane representing the road surface within each tire contact patch. For an uneven road, a different road plane may exist at each tire contact patch. For a typical road surface changing with a wavelength greater than several times the dimensions of the tire contact patch, a plane tangent to the road surface at the center of contact is an adequate representation of the road plane. When the individual road planes at each tire contact patch are essentially coplanar, the average lane can be taken to be the road plane. However, in the case of road surfaces with a wavelength similar to or less than the size of the contact patch, an equivalent road plane must be determined. This equivalent plane may not be coincident with the road surface at the tire contact center. This happens in the case of many ride events and the determination depends on the requirements of the analysis being performed. Very small wavelength, small amplitude, road surface undulations, such as the pebbles forming the road surface macrotexture, may be ignored in detail. For more information, see **J2047_201911** [15].

Road Rash

For a motorcycle rider, an abrasive skin injury usually caused by sliding on a rough hard surface, such as a paved roadway. For more information, see **J3133_SEP2019** [18].

Road Surface

The surface, flat, curved, undulated, or other shape, supporting the wheel (tire) and providing the friction necessary to generate tire shear forces in the road plane. See Road Plane. For more information, see **J2047_201911** [15].

Road User

A traffic participant on or adjacent to an active roadway for the purpose of traveling from one location to another. Road users include motor vehicles (including emergency vehicles), vehicle occupants, pedestrians, pedalcyclists, and users of motorized and nonmotorized personal mobility devices, such as scooters, wheelchairs, and mobility carts. Road users are governed by local traffic laws. For more information, see **J3216_JUL2021** [11].

Rocker Shifter

On a motorcycle, a foot-operated transmission gear-selector lever that allows the rider to shift gears by pushing down on front of the lever to downshift or down on the rear of the lever to upshift. For more information, see **J3133_SEP2019** [18].

Roll

See Pitch, Roll, Yaw; Vehicle Roll Angle. See Figure 42.

Rollbar

A structural member placed over the occupant compartment of a vehicle to protect the occupants against the effects of roof crush during vehicle rollover; also used in some buses and construction machinery.

Rolling Moment (Rolling Resistance Moment)

The moment exerted on the wheel by the road about the Y_T-axis. See Figure 4. For more information, see **J2047_201911** [15].

Rolling Radius

Distance in inches from center of tire to ground under load.

Rolling Resistance

The retarding force of a freely rolling wheel due to interaction with a contact surface, parallel to the heading axis of a wheel of a moving vehicle. Also

- A force opposite to the direction of travel resulting from deformation of a rolling tire. For more information, see **J670_JUN2022** [6] and **J1451**.
- Several resistances to motion that may be classified as due to friction in the wheel bearings, friction in the tire walls and tread as they flex when rolling along the road surface, deformation of road surface, impact resistance due to irregularities of road surface, and churning of air by wheels.

For more information, see **J2188_201807** [63].

Rollout

Part or all of a postimpact trajectory in which little or no sideslip of a vehicle's wheels occurs. See Spinout.

Rollover

Vehicle motion in which the wheels of the vehicle leave the road surface and at least one side or top of the vehicle contacts the ground. See Flip; Vault.

ROPS

Rollover protection system or rollover protection structure. A system or structure affixed to a moving vehicle that protects the belted occupant in the case of vehicle rollover.

ROR

Run off the road.

RPM

Revolutions per minute, generally of an automotive engine.

RTC

Real-time clock.

RV

Recreational vehicle.

RVC

Rear-view camera.

RWD

Rear-wheel drive.

S

SAD

Semi-autonomous driving; also secure autonomous driving, which protects autonomous vehicles from cyberattacks.

SAE Coordinate System

See Three-Axis Vehicle Coordinate System.

Safety Envelope

For a vehicle equipped with an automated driving system (ADS), a kinematically defined state space around a vehicle that represents a buffer between the subject vehicle and other objects in the environment. Maintaining a safety envelope contributes to a safer driving environment by allowing the ADS time and space to respond to the actions of nearby road users and other objects and by allowing other road users to potentially correct erroneous driving. For more information, see **AVSC00006202103** [61].

Safety Chain

An assembly that provides a secondary means of connection between the rear of the towing vehicle and the front of the trailer (or towed vehicle). For more information, see **J684_MAY2014** [40].

Safety Critical Event

Any conflict or series of related conflicts that involves the subject vehicle either alone or in combination with another vehicle, pedal cyclist, pedestrian, object, or road edge. For more information, see **J3151_201810** [65].

Safing (Arming) Sensor

A type of crash sensor that acts as a safety switch in a sensing system. A safing or arming sensor can be used as part of a distributed system to ensure the closure of a discriminating sensor is accompanied by another sensor. It can also be used in conjunction with electronic sensors to rule out electronic-magnetic interference as a cause of faulty sensor triggering. The general purpose of safing or arming sensors is to prevent inadvertent deployment of airbags. For more information, see *Fundamentals of Crash Sensing in Automotive Air Bag Systems* [64].

Sagittal Plane

The midplane dividing the human body into its left and right parts; also called the median plane. For more information, see *Principles of Biomechanics* [25].

SAI (SA)

Sudden acceleration incident (sudden acceleration).

S-Cam Brakes

The most common braking system on truck tractors and semitrailers that have pneumatic braking systems. S-cam braking systems are drum brake systems. The name derives from the profile of the cam that rotates, causing the brake shoes to rotate outward to engage with the drum that is rotating with the wheel, thus slowing the rotational rate of the tire and wheel assembly (see Figure 35).

FIGURE 35 Schematic of an S-cam drum brake system.

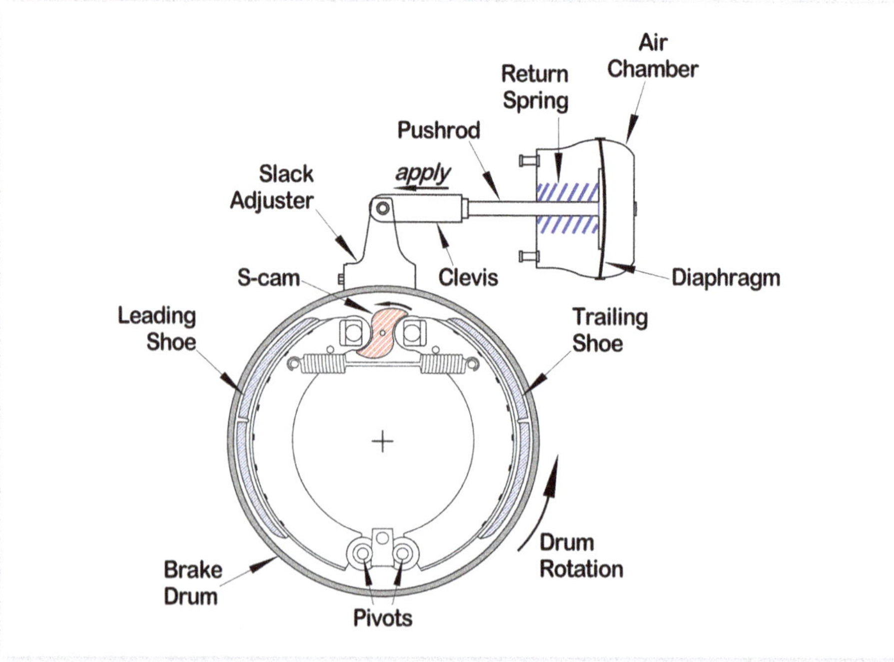

Scientific Method

As it applies to crash reconstruction, the scientific method involves gathering quantifiable, measurable evidence about a crash and testing a hypothesis about its causes and circumstances by applying Newtonian physics and using empirical data.

Scoliosis

Lateral curvature of the spine. For more information, see *Occupational* Biomechanics [19].

Scrape

A mark on a surface that is wider than it is deep that can usually be felt with fingers.

Scratch

A light and usually irregular scar made on a hard surface, such as paving, by a sliding metal part without great pressure. Scratches are visible but not normally distinguishable to the touch (see Figure 36).

FIGURE 36 Example of scuff marks (broad, dark, sweeping marks on the roadway surface) and gouge marks (light-colored marks into the surface).

Scuff Mark

Relatively short marks made by a moving tire on a road or other surface in an erratic fashion with no specific, consistent features, for example, acceleration scuff, impact scuff, flat tire mark (see Figure 36). See also Gouge; Skid Mark; Yaw Mark.

SDD

Sudden Deceleration Data (Cummins engines).

SDM

Sensing and diagnostic module. An electronic module in a vehicle that through preprogrammed algorithms and vehicle sensor inputs determines whether to deploy a vehicle's supplemental occupant restraints (e.g., airbags or seat belt pretensioners) in a crash-like event; General Motors' terminology for an airbag control module (ACM). The SDM may also capture and store vehicle event data before, during, or after a crash-like event in which supplemental occupant restraints may or may not have deployed. See ACM; Event Data Recorder; RCM.

Seat Belt Anchorage

Seat belt anchorage means any component, other than the webbing or straps, involved in transferring seat belt loads to the vehicle structure, including, but not limited to, the attachment hardware, seat frames, seat pedestals, the vehicle structure itself, and any part of the vehicle whose failure causes separation of the belt from the vehicle structure. For more information, see **J1803_JAN2013** [16].

Seat Belt Assembly

Any strap, webbing, or similar device designed to restrain an occupant in a motor vehicle during a crash. Includes all necessary buckles and other fasteners, and all hardware designed for installing such seat belt assembly in a motor vehicle. For more information, see **J1803_JAN2013** [16].

Seat System

The support and positioning mechanism for the occupant upon which the occupant sits. It may serve as the attachment point for seat belt hardware. It may include the seat back, seat cushion, seat adjusters, head rest, recliners, arm rests, and so on. For more information, see **J1803_JAN2013** [16].

Seating Reference Point (SgRP)

The unique design H-point (see Figure 26), as defined in SAE J1100, which has the following characteristics:

- Establishes the rearmost normal design driving or riding position of each designated seating position, which includes consideration of all modes of adjustment: horizontal, vertical, and tilt in a vehicle.
- Has X, Y, and Z coordinates, as defined in **J1100_JUN1984**, established relative to the designed vehicle structure.
- Simulates the position of the pivot center of the human torso and thigh.
- This is the reference point employed to position the two-dimensional drafting template with the 95th percentile leg described in **J826_May1987**, or if the drafting template with the 95th percentile leg cannot be positioned in the seating position, it is the most rearward adjustment of the seating position.

Secondary Control Adaptation

Provides the effective use of a motor vehicle's operating systems to a driver with a disability, so that he or she may drive and operate that motor vehicle with the same degree of safety as a nondisabled driver. Thus, the adaptive equipment must be

- Accessible to the driver with a disability for whom it is designed
- Not susceptible to inadvertent operation that may be inconvenient or dangerous for the driver and other users of the roadway
- Suitable for nondisabled drivers who may need to operate the motor.

For more information, see **J2388_202208** [66].

Second Impact

An impact between an occupant and an interior surface of a vehicle caused by and following an impact between the vehicle and another object.

Secondary Impact

A second or subsequent impact between the same two vehicles during a crash.

Section Height

Distance between the rim base and the surface of the tread for a vehicle tire. The unloaded section height at the "top" of the tire will be greater than the loaded section height being compressed against the pavement. For more information, see *Motor Truck Engineering Handbook* [7].

SEE

Search, evaluate, execute. SEE is a motorcycle riding strategy taught by the Motorcycle Safety Foundation (MSF) in their safety courses. For more information, see **J3133_SEP2019** [18].

Semi

Colloquial term for a tractor-trailer combination vehicle. See Tractor; Semitrailer.

Semitrailer

A semitrailer is a towed vehicle equipped with one or more axles to the rear of its laden center of gravity and whose front end forms part of a pivot joint attached to a truck tractor or other powered tow vehicle. Some examples of semitrailers include a dry box van carrying cargo, a recreational semitrailer (e.g., RV), a boat on a boat trailer, and livestock trailers.

"Semitrailer" means a trailer, except a pole trailer, so constructed that a substantial part of its weight rests upon or is carried by another motor vehicle. For more information, see 49 CFR §571.3 [43].

Separation

The time in the collision sequence when the collision forces have been dissipated and the objects in contact have the opportunity to physically separate from each other. For more information, see *Road Vehicle Dynamics* [8].

Separation Speed

The speed at the time of loss of contact of two vehicles in a collision; it can also refer to the speed of the centers of gravity or of the contact point.

Separation Velocity

The vector velocity at the time of loss of contact of two vehicles in a collision; it can also refer to the speed of the centers of gravity or at the contact point.

Service Brake, On and Off

The status of the device that is installed in or connected to the brake pedal system to detect whether the pedal was pressed. The device can include the brake pedal switch or other driver-operated service brake control.

Service Brake System

Primary brake system for stopping a vehicle. For more information, see *Motor Truck Engineering Handbook* [7].

SgRP

See H-Point; Seating Reference Point.

Shimmy

A self-excited, in-phase oscillation of a pair of steerable wheels about their steering axes, accompanied by appreciable tramp. For more information, see **J670_JUN2022** [6].

Shock Absorber

A damper that adds negligible kinematic constraint to the suspension (see Figure 37). For more information, see **J670_JUN2022** [6].

FIGURE 37 Shock absorber.

vershandrey/Shutterstock.com.

Shoulder Reference Point (SHRP)

A point 563 mm (22.16 in) above the design H-Point along the torso centerline of the SAE two-dimensional drafting template described in **J826**. This dimension, added to the dimension of 97.5 mm (3.84 in) from the H-Point to the buttocks flesh line at an angular relationship of 90 degrees between the torso and thigh segment of the two-dimensional drafting template, has been found to be representative of the shoulder height of the 99th percentile of the adult male driver population. For more information, see **J383_202207** [67].

SID

Subsystem identification (e.g., 10 is the ABS Axle 2 Right Pressure Modulation Valve SID). See SAE **J1587/J1708**. For more information, see *Commercial Vehicle Braking Systems: Air Brakes, ABS and Beyond* [5].

Side Airbag

Any inflatable occupant restraint device that is mounted to the seat or side structure of a vehicle's interior and that is designed to deploy in a side impact crash to help mitigate occupant injury and/or ejection.

Side Curtain/Tube Airbag

Any inflatable occupant restraint device that is mounted to the side structure of a vehicle's interior and is designed to deploy in a side impact or rollover crash to help mitigate occupant injury and/or ejection.

Side Rail

The outermost edge on the side of a vehicle's roof that connects the upper ends of the A, B, C, and D pillars. See Figure 1.

Side Stand

On a motorcycle, a retractable device that, when extended, supports a stationary motorcycle by providing a third contact point with the parking surface. Three-point ground support is provided by leaning the motorcycle toward the side stand until it is in contact with the ground (the stand and the motorcycle's two tires are the three contact points); more commonly called a kickstand.

Sideslip

Lateral/transverse translation of a vehicle perpendicular to its heading. See Figure 24 and Wheel Slip Angle.

Sideslip Angle, Tire

See Wheel Slip Angle.

Sideslip Angle, Vehicle

The angle between a vehicle's heading and its velocity vector (β in Figure 24). See Sideslip. The vehicle sideslip angle can be calculated from the longitudinal velocity, v_x, and the lateral velocity, v_y, using $\beta = \tan^{-1}(v_y/v_x)$. For more information, see **J670_JUN2022** [6].

Sideslip Coefficient

The slope of the initial linear portion of the lateral force-slip angle curve of a tire.

Sideslip Stiffness

See Cornering Stiffness; Sideslip Coefficient.

Sideswipe Collision

A collision of a vehicle or vehicles where sliding over the intervehicular contact surface (relative tangential motion) does not end at or before separation. See Common Velocity Conditions.

Sidewall

The portion of either side of the tire that connects the bead with the tread. For more information, see *Road Vehicle Dynamics* [8]. See Figure 7.

Simulation

The use of mathematics and mechanics, usually done using a computer, to represent, reproduce, or model a physical process.

SIPDE

An acronym for a safe motorcycle riding technique taught by the Motorcycle Safety Foundation (MSF) in their safety courses some years ago. It stands for Scan, Identify, Predict, Decide, and Execute. The MSF shortened that process to SEE: search, evaluate, execute. For more information, see **J3133_SEP2019** [18].

Sipe (Also Siping; Kerf)

A small slit or slits within the tread blocks. These are molded into the tire tread element at the time of cure and can be contained completely within the tread block or may exit the tread block and join the grooves or slots. The term "micro-siping" refers to small slits cut into the tread by a tire dealer. For more information, see *Tire Forensic Investigation* [26] and **J2047_201911** [15].

SIR

Supplemental inflatable restraint. A system, generally an airbag system, that supplements the fixed restraint system (seat belts) on a vehicle.

SI System of Units

Metric system of measurement (Système International d'Unitès).

Skid

The sliding of the entire tread contact area over the road surface. For more information, see **J2047_201911** [15].

Skid Number
A number representing tire-pavement frictional drag determined by measurements made according to standard equipment, conditions, and procedures and usually stated as 100 times a friction coefficient.

Skid Mark
A friction mark on a pavement made by a tire that is sliding without rotation and, if along the heading axis of the tire, can display a tread pattern.

Sliding Friction Coefficient
See Coefficient of Friction.

Slip
See Sideslip; Wheel Slip.

Slip Angle, Tire
See Wheel Sideslip Angle.

Slip Speed
The speed of a single wheel in the direction of its heading at a given value of longitudinal wheel slip.

Slip Stiffness
See Wheel Slip Coefficient.

SO2 (SO_2)
Sulfur dioxide.

SOC
State of charge (electric vehicles).

Soil Trip
A vehicle rollover that initiates as a result of the furrowing forces from the buildup of soil by the wheels as the vehicle moves laterally on a dirt surface. For more information, see **J2926_201707** [36].

Solenoid

A device that converts an electrical signal into mechanical movement. It consists of a coil with a moveable core that changes positions by means of electromagnetism when current flows through the coil. For more information, see *Commercial Vehicle Braking Systems: Air Brakes, ABS and Beyond* [5].

Solid-Axle Suspension

A suspension utilizing an essentially rigid axle. For more information, see **J670_JUN2022** [6]. See Independent Suspension.

Sound Level

Sound pressure level, $L_P = 10\log\frac{p^2}{p_{ref}^2}$, where p is the mean square acoustic pressure and $p_{ref} = 2 \times 10 - 5\,\text{N/m}^2$.

Sound power level, $L_W = 10\log_{10}\frac{W}{W_{ref}}$, where W is the acoustic power and $W_{ref} = 1 \times 10^{-6}W$.

SPA

Shoe plane angle. See H-Point. For more information, see **J826**, **J1100_NOV2009** [14].

Speed

The rate of change of vehicle displacement with respect to time; the magnitude of velocity.

Speed, Vehicle Indicated

The vehicle speed indicated by a manufacturer-designated subsystem designed to indicate the vehicle's ground travel speed during vehicle operation.

Sphenoid Bone

The frontal base of the skull. For more information, see *Principles of Biomechanics* [25].

SPI

Safety performance indicators. Used generally in aviation, an SPI is a parameter based on collected data that provides a metric pertaining to characteristics regarding incidents and accidents.

Spin

The limit condition for vehicle directional response wherein the rear tires have reached their cornering limit while the front tires have not reached their limit. See Drift; Plow. For more information, see J670_JUN2022 [6].

Spinout

A descriptive term for postimpact vehicle motion including significant yaw rotation. See Postimpact Trajectory.

Spring Brake

Emergency or auxiliary brake system utilizing a spring load as a force for braking. May be automatically actuated by low air pressure or mechanically controlled for use as a parking brake. For more information, see *Motor Truck Engineering Handbook* [7].

Sprung Mass

Includes body, frame, powertrain, payload, and driver. For more information, see *Motor Truck Engineering Handbook* [7].

Supination

Process of rotating the flexed forearm outward so that hand becomes "supine," that is, "palms up." For more information, see *Occupational Biomechanics* [19].

Stabilizer Bar (Anti-Roll Bar, Anti-Sway Bar)

A torsional spring that is loaded in a suspension roll to provide additional suspension roll stiffness. For more information, see **J670_JUN2022** [6].

Static Crush

See Residual Crush.

Static Roll Stability Limit (Static Rollover Threshold)

The maximum steady-state lateral acceleration a vehicle can sustain while remaining stable in roll. For more information, see **J670_JUN2022** [6].

Statics

The study of forces acting on a body in equilibrium.

Static Stability Factor (SSF)

The geometric stability ratio at a load condition consisting of curb weight plus a 50th percentile male (161.4 pound) driver. This load condition is used for evaluation of SSF in the US New Car Assessment Program (NCAP). For more information, see **J670_JUN2022** [6].

Steer Angle

The angle between the projection of a longitudinal axis of a vehicle and the line of intersection of the wheel plane and the road surface. For more information, see *Road Vehicle Dynamics* [8].

For each road wheel, the angle from the X_V axis to the wheel plane, about the Z_V axis. See angle δ in Figure 24. For more information, see **J670_JUN2022** [6].

Steering Axis (Kingpin Axis)

The axis of rotation of the knuckle relative to a vehicle's sprung mass when steered. For more information, see **J670_JUN2022** [6].

Steering Gear

A gear train and/or mechanism, with housing, carrying steering motion and loads from the intermediate shaft to the tie rods or Pitman arm. A steering gear may have provisions for power assist. For more information, see **J670_JUN2022** [6].

Steering Input

The angular displacement of the steering wheel measured from the straight-ahead position (position corresponding to zero average steer angle of a pair of steered wheels).

Steering Ratio

The rate of change of steering-wheel angle with respect to the mean reference steer angle of a pair of steered wheels at a given steering-wheel position. For more information, see **J670_JUN2022** [6].

Steering Wheel Angle

The angular displacement of the steering wheel measured from a predetermined straight-ahead position. For more information, see **J670_JUN2022** [6].

Stiffness Coefficient

See Crush Stiffness Coefficient.

Stoppie

For a motorcycle, a large amplitude, pitch-forward condition whereby the rear wheel lifts off the ground; usually caused by a combination of extreme rider front wheel braking and body movement fore and aft. Also known as a front wheel stand (see Figure 38). For more information, see **J3133_SEP2019** [18].

FIGURE 38 Motorcycle and rider doing a stoppie.

Goga Shutter/Shutterstock.com.

Stopping Distance

The distance taken by a driver to bring a vehicle to rest in straight forward motion by braking, including the distance traveled during perception-decision-reaction time prior to brake application. See Braking Distance.

STRAHNET

Strategic Highway Network. The network of highways, including the US Interstate System, that provides connecting routes to military installations, industries, and resources.

Straight Free-Rolling Tire

A loaded-rolling tire moving without applied torque along a linear path (zero path curvature) at zero slip and inclination angles. Note: In defining the straight free-rolling tire, it is assumed that (a) the tire rotates at a constant spin velocity and (b) the tire is pulled or pushed by applying a longitudinal force at its center, sufficient to balance rolling loss and friction torque. For more information, see **J2047_201911** [15].

Strain

Deformation (change in dimension) that develops within a structure in response to externally applied loads expressed as the change in dimension per unit length, for example, in./in.

Stress

Load per unit area that develops on a plane surface within a structure in response to externally applied loads. For more information, see *Basic Biomechanics of the Musculoskeletal System* [3].

Striations

Periodic stripes that appear transverse to the tire marks from a yawing vehicle (see Figure 39).

FIGURE 39 Striations in a tire mark (SAE 2009-01-0092).

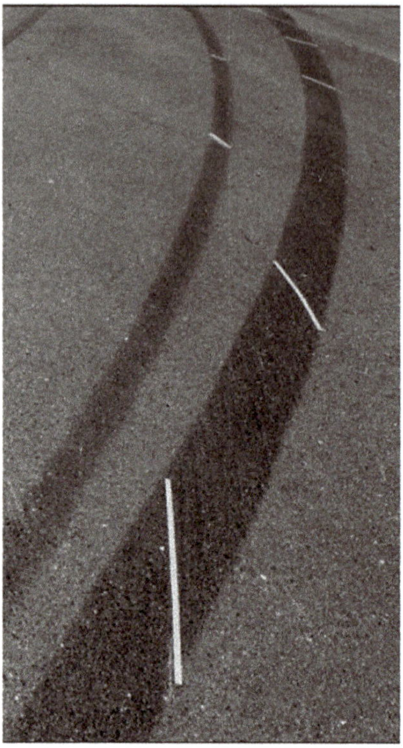

Strut

A damper that also contributes to rotational constraint of the knuckle in camber and caster (see Figure 40). For more information, see **J670_JUN2022** [6].

FIGURE 40 A pair of strut suspension elements.

Stud-Mount Wheels

Wheels that are designed to center on the studs of a hub. These wheels have chamfers at the bolt holes into which a ball seat or conical nut is installed to center the wheel. The center bore of the wheel is only for clearance of the axle end. For more information, see *Motor Truck Engineering Handbook* [7].

Subcutaneous

Beneath the dermis (skin). For more information, see *Principles of Biomechanics* [25].

Subdural Hematoma

Bleeding between the two layers surrounding the brain. For more information, see **J885_FEB2011** [4].

Submarining

The phenomena where the anterior superior iliac spine (ASIS) of the pelvis slides beneath the lap portion of a properly worn seat belt during a crash. This can result in compression of the soft portion of the occupant's abdomen. For more information, see **J383_202207** [67].

Supercharger

For an internal combustion engine, a device that, using a direct connection to the engine, drives a centrifugal blower to force greater mass of air into the engine cylinders so that the engine can produce more power.

Superelevation

A side-to-side slope of a road measured in degrees or percent; also referred to as a banked roadway.

Super Single Tire

A single, wide tire and wheel combination for use on motor homes and commercial vehicles, typically 15 inches (385 mm) wide, or wider, that provides the same load-carrying capacity as a dual tire and wheel combination; also called a "wide base" tire.

Supine

Lying on the back. For more information, see *Principles of Biomechanics* [25].

Supplemental Restraint System (SRS)

A system of vehicle interior components intended to enhance the restraint of the vehicle's occupants provided by the vehicle's seat belts and reduce the risk of injuries to the occupants in various crash scenarios. SRS typically refers to a suite of airbags in combination with seat belt pretensioners.

Suspension

A system that provides ride freedom and kinematic control of the motions of the wheels at a given longitudinal position on a vehicle. See Independent Suspension. For more information, see **J670_JUN2022** [6].

SUV

Sport utility vehicle. A light truck designed to carry passengers and cargo within a single, continuous, enclosed occupant compartment and intended for on-highway and limited off-highway use. SUVs are distinguished by being derived from pickup truck designs and incorporating body-on-frame architecture as opposed to crossover utility vehicles (CUVs) that are derived from passenger car designs and incorporate unibody architecture.

SVC

Surround view camera. A camera, or system of cameras, that provides panoramic view(s) that assist the driver in parking a vehicle.

System Failure (DDT Performance Relevant)

A malfunction in a driving automation system and/or other vehicle system that prevents the driving automation system from reliably performing the portion of the dynamic driving task (DDT) on a sustained basis, including the complete DDT, that it would otherwise perform. This definition applies to vehicle fault conditions and driving automation system failures that prevent a driving automation system from performing at full capability according to design intention. For more information, see **J3016_JUN2018** [9].

T

Tag Axle

See Lift Axle.

Tandem Axles

The use of two closely spaced axles, front-to-rear, usually for buses, heavy trucks, truck tractors, and trailers (see Figure 41).

FIGURE 41 Tandem axles on a truck tractor.

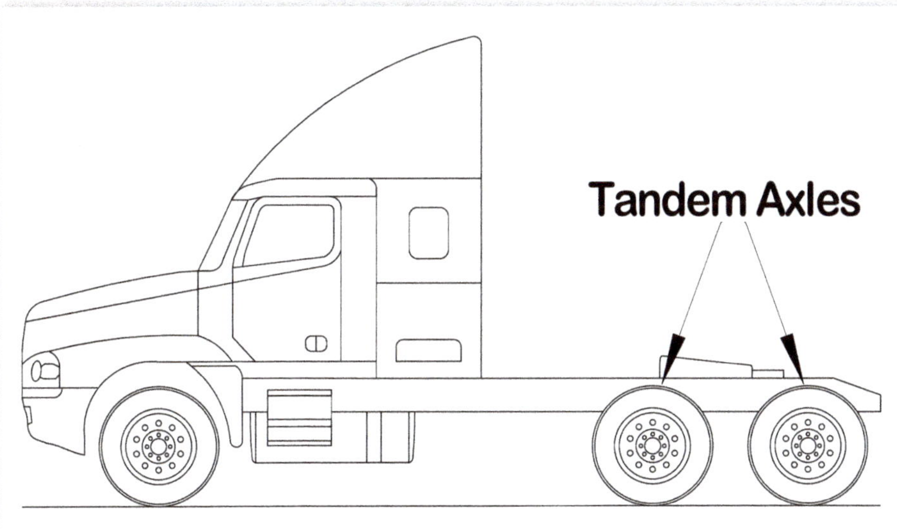

Tare Weight

Weight of an empty vehicle or container.

Target Fixation

For a motorcycle rider, when the rider's eyes focus or "fix" onto a point or object in the road or trail ahead, such as debris or a rock, the rider may inadvertently steer the bike toward that "target" rather than in the intended path. The condition where a motorcycle rider is visually and mentally focused on one object to the exclusion of everything else. For more information, see **J3133_SEP2019** [18].

TAU

Throttle actuation unit.

TCS

Traction control system. A system that detects a loss of traction at one (or more) wheels and can automatically apply the brake(s) or reduce the engine torque directed to the affected wheel(s).

TEA-21

Transportation Equity Act for the 21st Century.

Thorax

The area of the body between the neck and the abdomen.

Three-Axis Vehicle Coordinate System

Figure 42 shows the standard, three-dimensional vehicle coordinate system. For more information, see **J670_JUN2022** [6].

FIGURE 42 Diagram of a three-axis SAE coordinate system: (A) System with Z upward, (B) System with Z downward.

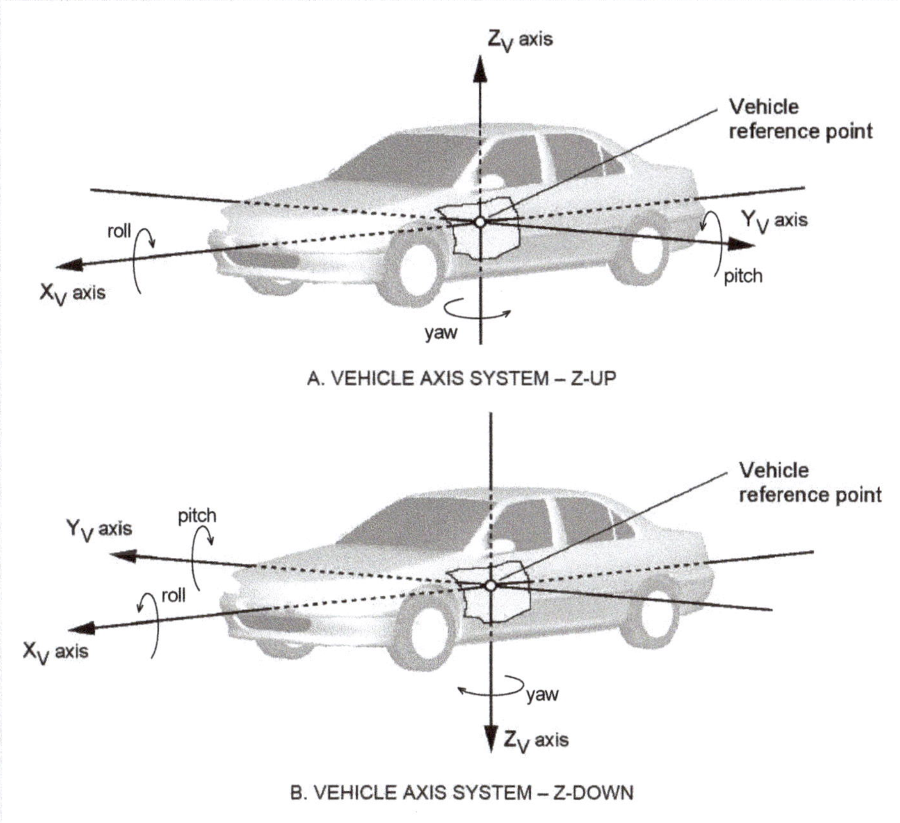

3-Point Harness

A harness found only on infant restraints that consists of shoulder straps coming together at a buckle in the shell or on the crotch strap; not to be confused with 3-point (lap-shoulder) belts. For more information, see **J2939_201304** [2].

Throw Distance

The distance a pedestrian is propelled (in the direction of vehicle motion at impact) between the location of the pedestrian at first contact and the pedestrian's rest position (see Figure 43).

FIGURE 43 Pedestrian throw distance (wrap collision shown).

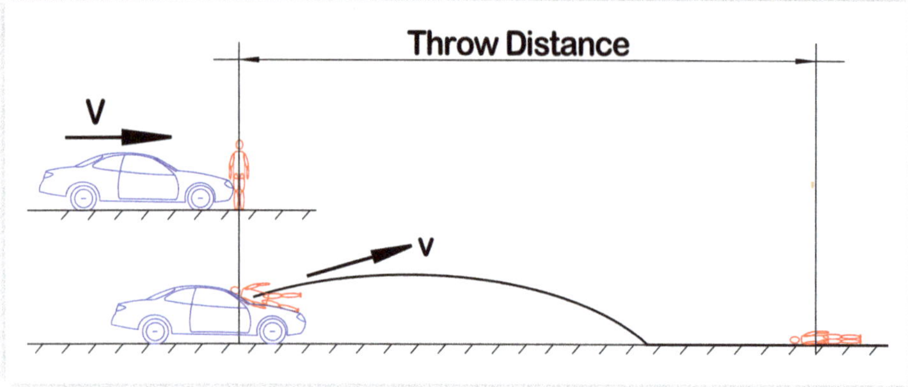

TIFF (TIF)

Tagged Image File Format; a digital photograph file format.

Time (Maximum Delta-V, Longitudinal)

For an event data recorder (EDR), the time from crash time zero to the point where the maximum value of the cumulative change in velocity is found, as recorded by the EDR, along the longitudinal axis. For more information, see **J1698_MAR2023** [41].

Time (Maximum Delta-V, Resultant)

For an event data recorder (EDR), the time from crash time zero to the time where the maximum delta-V resultant occurs, as recorded by the EDR or processed during data download. For more information, see **J1698_MAR2023** [41].

Time Duration of Impact Event

For an event data recorder (EDR), the duration of an impact event is the time interval between the beginning of an impact event (time zero) and the end of the impact event (T_{END}). Duration of an impact event is not a predetermined time interval and can vary significantly for each event. The duration of an impact event may exceed the time recording capability of the EDR. For more information, see **J1698_MAR2023** [41].

Time End (End of Impact Event)

For an event data recorder (EDR), the end of an impact event (T_{END}) depends on the type of occupant protection control algorithm:

- For "wake-up" occupant protection control algorithms, T_{END} occurs at the moment when the occupant protection control algorithm resets itself. If this condition has not been met by the end of the recording period, T_{END} may be defined as the last recorded data point.
- For continuously running occupant protection control algorithms, T_{END} occurs when the longitudinal and lateral (if recorded) cumulative delta-V, within a 20-ms time period, becomes 0.8 km/h (0.5 mph) or less. If this condition has not been met by the end of the recording period, T_{END} may be defined as the last recorded data point.

For more information, see **J1698_MAR2023** [41].

Time Zero (Beginning of an Impact Event)

For an event data recorder (EDR), the beginning of an impact event (time zero) is defined by the following:

- For systems with "wake-up" occupant protection control algorithms, it is the time at which the occupant protection control algorithm is activated.
- For continuously running occupant protection control algorithms, it is the time when the cumulative delta-V of over 0.8 km/h (0.5 mph) is reached within a 20-ms time period in the longitudinal direction for a frontal/rear event or within a 5-ms time period in the lateral direction for a side impact event.

Or

- A deployment of a nonreversible deployable protection device occurs. This does not include the deployment of subsequent stages of a multistage deployable protection device.

The beginning of an impact event should not be interpreted as the moment when the vehicle makes a first physical contact with another object. The beginning of an event also does not mark the condition that determines whether to record the event data associated with the event for subsequent retrieval. Instead, the purpose of defining the beginning of an event is to enable the alignment of data elements from multiple recording devices on a vehicle. For more information, see **J1698_MAR2023** [41].

Tire

See Pneumatic Tire.

Tire and Wheel Coordinate System (Z-Down)

A reference system intended for the orientation of forces and moments exerted on a tire by the road and also for angular orientation of the tire with respect to the road plane. It is an orthogonal, right-handed, three-axis coordinate system originating at the contact center. The X_T and Y_T axes are located in the road plane. The Z_T axis is perpendicular to the road plane and is positive into-the-road plane. The X_T axis coincides with the contact line. It is positive forward. The Y_T axis is positive to the right (see Figure 44). For more information, see **J2047_201911** [15].

FIGURE 44 Tire and wheel coordinate system (Z-down).

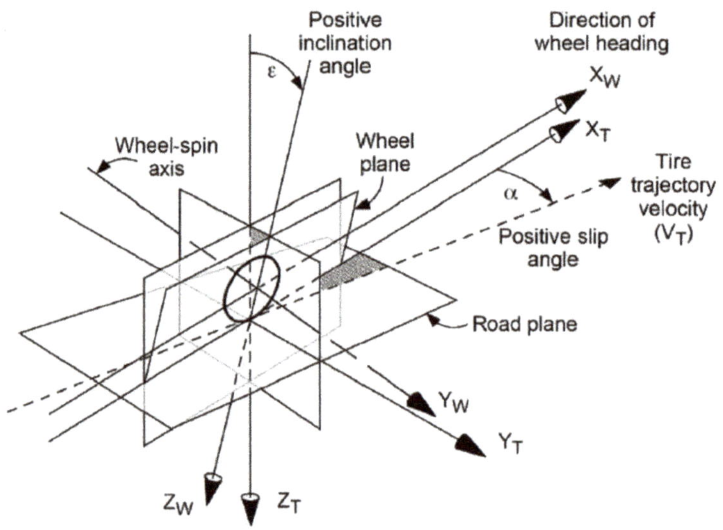

Tire Aspect Ratio

Ratio of the section height to the section width of a tire times 100. For more information, see **J2047_201911** [15].

Tire Belt (Breaker)

An assembly of plies located under the tread that does not extend into the sidewalls. It provides additional tread area stiffness/strength. For radial ply constructions, it restrains the overall diameter, provides circumferential tread stiffness, and is the source of cornering forces. For more information, see **J2047_201911** [15].

Tire Carcass (Body; Casing)

The rubber-bonded cord structure that provides a tire's stiffness when prestressed by the inflation pressure. For more information, see **J2047_201911** [15].

Tire Cord

An assembly formed by twisting together textile or non-textile filaments that is the structural reinforcing element for plies. For more information, see **J2047_201911** [15].

Tire Deflection

The amount the tire section height is reduced due to tire normal force. For more information, see **J670_JUN2022** [6].

Tire DOT Code

The DOT code is comprised of the symbol DOT and the tire identification number (TIN). The TIN is comprised of letters and numbers molded or branded into or onto the sidewall of the tire. For example, for the code DOT YY XX ZZZZ DODD, YY indicates the tire plant, XX indicates the tire size, ZZZZ is an optional code, and DODD represents the week and year of manufacture (4007 would indicate the 40th week of 2007). See 49 Code of Federal Regulations part 574 for details. For more information, see **J2047_201911** [15].

Tire Footprint (Contact Patch)

The contact area of a tire loaded against a flat or curved surface. For more information, see **J2047_201911** [15].

Tire Force

A vector quantity expressing the sum of the forces exerted on a tire by the road at any instant, with its line of action passing through the contact center. See Figure 4. For more information, see **J2047_201911** [15].

Tire Groove

A void that is molded or cut into the tread rubber and is relatively narrow compared to its length. Sipes and kerfs are not grooves. For more information, see **J2047_201911** [15].

Tire Inflation Pressure

The gauge pressure of the filling gas within a tire under usage conditions. For more information, see **J2047_201911** [15].

Tire Inner Liner

A low air diffusion layer covering the inside of the carcass of a tubeless tire. For more information, see **J2047_201911** [15].

Tire Load Index

A numerical code associated with the maximum load a tire can carry at the speed indicated by its speed symbol under specified service conditions. For more information, see **J2047_201911** [15].

Tire Lower Sidewall

The lower sidewall is the sidewall region between the point of midsection height and the bead. For more information, see **J2047_201911** [15].

Tire Lug or Block

A discontinuous tread element. For more information, see **J2047_201911** [15].

Tire Marks

General term for marks on a surface generated by tires; can be scuffs, skids, yaw marks, prints, among others.

Tire Moment

A vector quantity expressing the sum of the moments exerted on the tire by the road at any instant, consistent with the line of action of the tire force. For more information, see **J2047_201911** [15].

Tire Overall (Outside) Diameter

The diameter of the largest part of the unloaded new tire mounted on the test rim and inflated to the recommended pressure including 24-hour inflation growth. For more information, see **J2047_201911** [15].

Tire Overall Width

The width of an unloaded new tire inflated to recommended pressure 24 hours prior to measurement, thus taking account inflation growth. Molded on sidewall elements such as curb ribs, lettering, and decorations are included. For more information, see **J2047_201911** [15].

Tire Ply

A sheet of rubber-coated cords. For more information, see **J2047_201911** [15].

Tire Section Height

The height of the radial cross section of a tire, including 24-hour inflation growth. It is calculated as half the difference between the tire overall diameter and the nominal rim diameter. For more information, see **J2047_201911** [15].

Tire Section Width

The width of an unloaded new tire mounted on a specified rim, inflated to the normal recommended pressure 24 hours prior to measurement thus taking account inflation growth, and including the normal sidewalls, but excluding molded-on sidewall elements, such as the protective rib, bars, and decorations. For more information, see *Road Vehicle Dynamics* [8] and **J2047_201911** [15].

Tire Service Description (Load/Speed Index)

A code consisting of load index and speed symbol, which is not part of the tire size designation, for example, 90 H. For more information, see **J2047_201911** [15].

Tire Sidewall

The portion of the tire between the bead and the tread (see Figure 7). See Tire Lower Sidewall; Tire Upper Sidewall. For more information, see **J2047_201911** [15].

Tire Size Designation (for Metric Passenger Car and Light Truck Tires)

The numbers or letters indicating intended tire application, nominal section width, nominal aspect ratio, construction, and nominal rim diameter, such as P205/60R17. Based on the *Tire and Rim Association Year Book*, P205/60R17 90H denotes a passenger car tire (P) with nominal section width of 205 mm, nominal aspect ratio of 60%, radial ply construction (R), with nominal rim diameter of 17 inches, load index 90, and speed symbol H. For more information, see **J2047_201911** [15].

Tire Slip Angle (Also Tire Sideslip Angle)

See Wheel Slip Angle.

Tire Speed Symbol

A symbol indicating the speed category at which the tire can carry a load corresponding to its load index under specified service conditions. For more information, see **J2047_201911** [15].

Tire Splice

The joint formed by overlapping or butting the ends of a tire component during tire building. For more information, see **J2047_201911** [15].

Tire Structure (Construction)

The generic type of tire structure is identified by a letter code. The letter "D" is used for diagonal tire, the letter "B" for bias belted tire, and the letter "R" for radial tire. For more information, see **J2047_201911** [15].

Tire Tread

The portion of the tire designed to contact the road surface in normal service. See Figure 7. For more information, see **J2047_201911** [15].

Tire Tread Crown

The crown is the region between the tread shoulders of the tire when looking from the inside center of the tire to the tire's outside. For more information, see **J2047_201911** [15].

Tire Tread Element (Projection)

A raised portion of the tread pattern, contacting the road surface when passing through the footprint. For more information, see **J2047_201911** [15].

Tire Tread Pattern

The molded geometric configuration on the peripheral tread face, generally composed of tread elements and voids. For more information, see **J2047_201911** [15].

Tire Tread Shoulder

The outermost portion of the tread adjacent to the sidewall. For more information, see **J2047_201911** [15].

Tire Tread Wear Indicator (Wear Bar)

Raised bottom portions of a groove or void, spaced regularly around the tire across the tread to provide a visual indication of wear-out. For more information, see **J2047_201911** [15].

Tire Upper Sidewall

The upper sidewall is the sidewall region between the point of midsection height and the tread. For more information, see **J2047_201911** [15].

Tire Void

An open space between tread elements or ribs. Sometimes used to describe the volume of the open space. For more information, see **J2047_201911** [15].

TLR

Traffic light recognition. The function of a system that detects the status of roadway traffic lights.

Toe-In/Toe-Out

A condition for the steer wheels of a vehicle whereby the fronts of the wheels are closer together or farther apart than the rears of the wheels (see Figure 45).

FIGURE 45 Diagram showing toe-in and toe-out on a light vehicle.

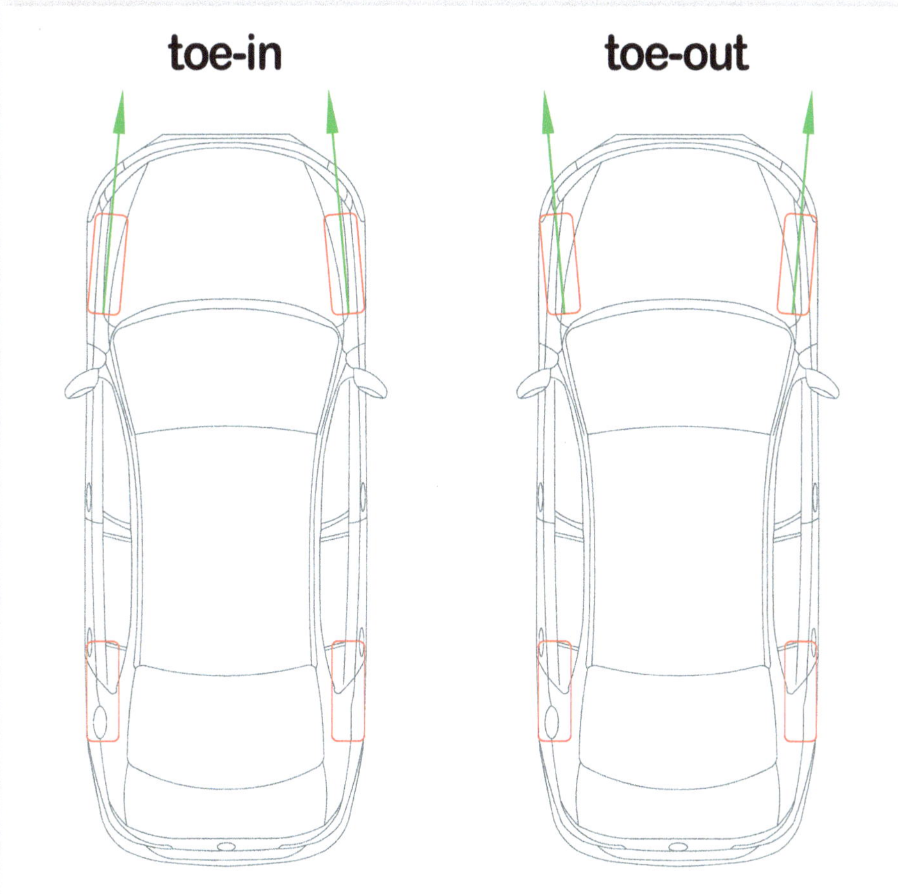

Tolerance Level

In biomechanics, the magnitude of loading that produces a specific injury level. For more information, see **J885_FEB2011** [4].

Tone Wheel

Another term used to refer to a tooth wheel in an antilock brake system (ABS) system. See Tooth Wheel. For more information, see *Commercial Vehicle Braking Systems: Air Brakes, ABS and Beyond* [5].

Tooth Wheel

A metal ring with evenly spaced teeth. It is usually attached to the hub at the wheel end on each antilock brake system (ABS)–monitored wheel. When the wheel rotates, the teeth move past the wheel speed sensor to create an electrical signal that the electronic control unit (ECU) uses to determine wheel speed. The ring will normally have 100 evenly spaced teeth, though it could have 80 or 120 teeth on extremely small or large tires. For more information, see *Commercial Vehicle Braking Systems: Air Brakes, ABS and Beyond* [5].

Top Tether

An assembly consisting of a top tether strap, attached at or near the top of a child restraint, together with a length adjuster and a top tether hook, that is used with either a vehicle belt or lower LATCH attachments for complete installation of a child restraint in a vehicle. For more information, see **J2939_201304** [2].

Total Station

An electro-optical, laser-based measuring device, usually mounted on a tripod, used to make precise measurements of bearing, elevation, and distance from the head unit to points of interest, such as in site surveys.

Track Width

The distance between the center of the tire tread on one side of a vehicle and the center of the corresponding tire tread on the opposite side. For dual-tired axles, the track width is measured between the centers of each dual wheel assembly of a given axle.

Tractor

See Farm Tractor; Truck Tractor.

Tractor Protection Valve

A valve that protects the air supply on the truck tractor from being depleted in the event of a disconnect of the service brake gladhand between the tractor and the trailer.

Tractor, Semitrailer (also Tractor-Semitrailer)

A truck tractor with two or more axles pulling a semitrailer (see Figure 46).

FIGURE 46 Illustration of a truck tractor with attached semitrailer.

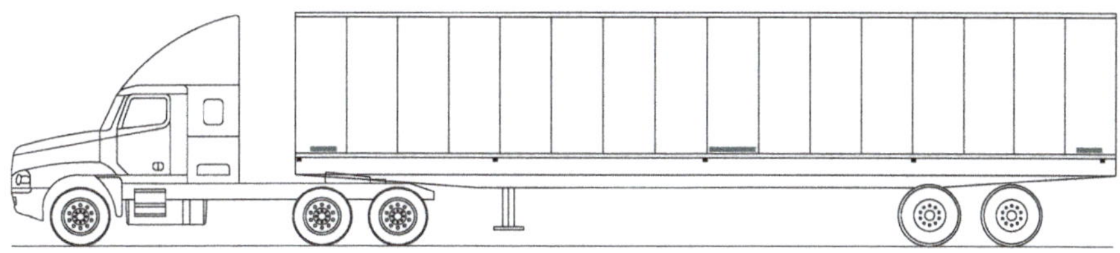

Tractor Trailer

A truck tractor with two or more axles pulling a full trailer or semitrailer.

Traffic Participant

Entities whose actions influence travel in the transportation environment, which may include road users engaged in travel upon or across publicly accessible roadways and road operators. For more information, see **J3216_JUL2021** [11].

Trailer

A trailer is a towed vehicle equipped with two axles; the front axle is attached to the tow vehicle and pivoted for turning, whereas the rear axle is fixed. Trailer also means a motor vehicle with or without motive power, designed for carrying persons or property and for being drawn by another motor vehicle. For more information, see 49 CFR §571.3 [43].

The definition of trailer includes the following:

- Full trailer means any motor vehicle other than a pole trailer that is designed to be drawn by another motor vehicle and so constructed that no part of its weight, except for the towing device, rests upon the self-propelled towing motor vehicle. A semitrailer equipped with an auxiliary front axle (converter dolly) is considered a full trailer.
- Pole trailer means any motor vehicle that is designed to be drawn by another motor vehicle and attached to the towing motor vehicle by means

of a "reach" or "pole," or by being "boomed" or otherwise secured to the towing motor vehicle, for transporting long or irregularly shaped loads such as poles, pipes, or structural members, which generally are capable of sustaining themselves as beams between the supporting connections.

- Semitrailer means any motor vehicle, other than a pole trailer, that is designed to be drawn by another motor vehicle and is constructed so that some part of its weight rests upon the self-propelled towing motor vehicle. For more information, see 49 CFR §390.5 [44]

Trailer Assistance

Assists a driver with visual guidance while backing toward a trailer or during backing maneuvers with a trailer attached. Some systems provide additional images while driving or backing with a trailer. Some systems provide steering assistance during backing maneuvers. For more information, see **J3063_MAR2021** [23].

Trailing Edge

The term used to describe that portion of a vehicle component (such as door, window, fender, quarter, etc.) which is closest to the rear of the vehicle. The rearmost part of a vehicle with respect to its motion and attitude.

Trajectory

The path of the center of gravity of a body as it moves through space; usually associated with coordinates of the center of gravity as a function of time. See Figure 24.

Tramp

The form of wheel hop in which the wheels on an axle hop in opposite phase. For more information, see **J670_JUN2022** [6].

Transaxle

A device that combines a transmission with a final drive unit which includes a differential. For more information, see **J1715_SEP2022** [17].

Transfer Case

Split power gearbox transmitting drive to front and rear axles. For more information, see *Motor Truck Engineering Handbook* [7].

Transmission
A device for transmitting power at a multiplicity of speed and torque ratios. For more information, see **J1715_SEP2022** [17].

TRB
Transportation Research Board. The TRB is part of the National Academies of Sciences, Engineering, and Medicine. The TRB provides independent information to governmental agencies, thereby promoting exchange of transportation-related information and research regarding all modes of transportation.

Tread Depth
The distance between the base of a tire tread groove and a line tangent to the surface of the two adjacent tread ribs or rows. For more information, see *Road Vehicle Dynamics* [8].

Tread Pattern
The molded configuration on the face of the tread, generally composed of ribs, rows, grooves, bars, lugs, and the like. For more information, see *Road Vehicle Dynamics* [8].

Trigger Threshold
For light vehicle event data recorders (EDRs), a change in vehicle velocity, in the longitudinal direction, that equals or exceeds 8 km/h within a 150-ms interval. For vehicles that record "delta–V, lateral," trigger threshold means a change in vehicle velocity in either the longitudinal or lateral direction that equals or exceeds 8 km/h within a 150-ms interval.

Trim Height
A vertical dimension that specifies the location of a fixed point on a vehicle body or chassis relative to ground. For more information, see **J670_JUN2022** [6].

Trip Point
That location along a ground surface at which the motion of a vehicle component is suddenly halted followed by a flip, rollover, or vault.

Trochanter
Either of the two bony protuberances at the upper part of the femur.

Truck

A motor vehicle with motive power, except a trailer, designed primarily for the transportation of property or special purpose equipment. For more information, see 49 CFR §571.3 [49].

Any self-propelled commercial motor vehicle except a truck tractor, designed and/or used for the transportation of property. For more information, see 49 CFR §390.5 [44].

Truck Classifications

Heavy trucks are classified according to their GVWR:

Class	GVWR range, lb	Class	GVWR range, lb
1	0-6,000	2	6,001-10,000
3	10,001-14,000	4	14,001-16,000
5	16,001-19,500	6	19,501-26,000
7	26,001-33,000	8	33,001 and above

Light-duty vehicles have a GVWR of less than 10,000 pounds; medium-duty vehicles have a GVWR of 10,001–26,000 pounds; heavy-duty vehicles have a GVWR greater than 26,001 pounds.

Truck Deformation Classification (TDC)

A classification system used to appropriately describe a collision-damaged truck. It consists of seven alphanumeric characters arranged in specific order to form a descriptive composite of the vehicle damage. For more information, see **J1301_202206** [68].

Truck Tractor

A motor vehicle designed for pulling semitrailers through a fifth wheel located near the rear axle(s). Basic types are cab over engine and conventional (see Figure 47).

- Truck tractor means a self-propelled commercial motor vehicle designed and/or used primarily for drawing other vehicles. For more information, see 49 CFR §390.5 [44].
- Truck tractor means a truck designed primarily for drawing other motor vehicles and not constructed to carry a load other than a part of the weight of the vehicle and the load drawn. For more information, see 49 CFR §571.3 [43].

- Truck tractor has the meaning given in 49 CFR §571.3 and 49 CFR §535.5(c). This includes most heavy-duty vehicles specifically designed for the primary purpose of pulling trailers but does not include vehicles designed to carry other loads. For purposes of this definition "other loads" would not include loads carried in the cab, sleeper compartment, or toolboxes. Examples of vehicles that are similar to tractors but that are not tractors under this part include dromedary tractors, automobile haulers, straight trucks with trailers hitches, and tow trucks. For more information, see 49 CFR §523.2 [45].

FIGURE 47 Truck tractors.

TSA

Transportation Security Administration. Founded as part of the Aviation and Transportation Security Act of 2001, the TSA is an agency of the US Department of Homeland Security whose mission is to protect and secure the nation's transportation systems and ensure freedom of movement of people and goods.

TSB

Technical Service Bulletin. A TSB is a document issued by a vehicle manufacturer that describes a known, non-safety-related issue with a vehicle and typically provides information pertaining to identification of the issue and corrective action, as well as recommended procedures for repairing of the issue.

TSC

Transportation Systems Center (NHTSA).

TSR

Traffic sign recognition. An electronic system that detects and recognizes roadway traffic signs and provides this information to a driver via an instrument cluster or other means.

TTC

Time to collision

Tuck-In

Transient roll phenomenon that may occur when the rider closes the throttle or declutches while the motorcycle is turning. For more information, see **J3133_SEP2019** [18].

Turbocharger

For an internal combustion engine, a turbine that uses the energy from exhaust gases to drive a centrifugal blower to force greater mass of air into the engine cylinders. For more information, see *Motor Truck Engineering Handbook* [7].

Turning Radius

The radius of the smallest circular path a vehicle can make with the steering system at full lock, typically scribed by the point on the vehicle farthest from the center of rotation. It is a function of the wheelbase, the front tire width, and the angle the front wheels can be steered.

U

UA

Unintended acceleration. Also sudden acceleration.

Underride

A condition in a collision where the main structural components of one vehicle are below the main structural components of the other vehicle. See Figure 32. Also see Override.

Understeer

Understeer occurs when a vehicle traveling along a circular path on a flat, level surface with a constant speed and a constant front-wheel steer angle is accelerated and it tends to increase its radius. See Neutral Steer; Oversteer.

Uniform Acceleration

Acceleration that continues over time at a constant magnitude.

Unladen Weight

The maximum weight of a vehicle (or container) with no payload. Also known as the tare weight of the vehicle (or container).

Unlocked Event

An event data recorder (EDR) record that may be overwritten by subsequent events. For more information, see **J1698_MAR2023** [41].

Unsprung Mass

Vehicle mass not supported by the suspension springs, such as tire, wheel, and axle assembly.

Unstabilized Situation

A set of events not under human control. It originates when control is lost and terminates when control is regained or, in the absence of persons who are able to regain control, when all persons and property are at rest. For more information, see **J3151_201810** [65].

Untripped Rollover

A rollover that initiates on the road surface as a result of friction forces between the tires and the road surface. For more information, see **J2926_201707** [36].

User

In the context of automated driving systems, a general term referencing the human role in driving automation. The following four terms describe categories of (human) users:

1. Driver
2. Passenger
3. Dynamic driving task (DT) fallback-ready user
4. Driverless operation dispatcher

For more information, see **J3016_JUN2018** [9].

User Monitoring and Prompting System (UMAPS)

A type of countermeasure that utilizes a user monitoring system to measure some aspect of user behavior, which is compared with a threshold value on that behavior measurement, and, when exceeded, provides a prompt designed to correct the user behavior. For more information, see **J3114_DEC2016** [32].

USGS

United States Geological Survey. The USGS is an agency within the US Department of Interior that provides the government and the public with science-based information regarding land use, ecosystems, energy and mineral resources, natural hazards, and water use. Additionally, the USGS provides maps and images of the Earth available to the public.

UST

Underground storage tank.

V

V2H
Vehicle to home (electronic communication).

V2I
Vehicle to infrastructure (electronic communication).

V2V
Vehicle to vehicle (electronic communication).

Valve Hole
The hole or slot in the rim that accommodates the valve for tire inflation. See item 17 in Figure 19. For more information, see **J2047_201911** [15].

Vault
The roll or pitch motion a vehicle makes following loss of ground contact. See Flip.

VCE
Vehicle control electronics.

Vehicle
A machine designed to provide conveyance on public streets, roads, and highways. For more information, see **J3016_JUN2018** [9].

Vehicle Coordinate System
See Figure 24 and Figure 42.

Vehicle Identification Number (VIN)

Since 1981 in the North American light vehicle and heavy vehicle markets, the 17-digit number assigned to every vehicle that uniquely identifies that vehicle.

Vehicle Length

The maximum dimension measured longitudinally between the foremost point and the rearmost point in a vehicle, including bumper, bumper guards, tow hooks, and/or rub strips if standard equipment. For more information, see **J1100_NOV2009** [14]. Also known as Overall Length (OAL).

Vehicle Width

The maximum dimension measured between the widest point on a vehicle, excluding exterior mirrors, flexible mud flaps, and marker lamps, but including bumpers, moldings, sheet metal protrusions, or dual wheels if standard equipment. For more information, see **J1100_NOV2009** [14]. Also known as overall width (OAW).

Vehicle Roll Angle

The angle between a vehicle's *y*-axis and the ground plane.

Velocity

The rate of change of displacement with both a magnitude and direction, and therefore is a vector quantity; the magnitude of velocity is referred to as speed.

Velocity-Time Curve

A graphical depiction of velocity of the center of gravity of a vehicle as it changes over time: $\boldsymbol{v} - \boldsymbol{\tau}$ or v - t curve.

Viewing Angle

The angle between an observer's line of sight when looking straight ahead and the line of sight when viewing a particular object (or area); it is the extent the eyes or head must rotate to look directly at that object. For more information, see **J264_OCT2021** [34].

VIIC

Vehicle Infrastructure Integration Consortium.

Visual Adaptation

A change in visual sensitivity as a result of continued or repeated exposure to stimuli of similar nature or magnitude. Adaptation may result in increased sensitivity (as is the case with dark adaptation) or in decreased sensitivity (which can result in aftereffects, such as successive color contrast). For more information, see **J264_OCT2021** [34].

Visual Distraction

Any glance that competes with activities necessary for safe driving. For more information, see **J3077_DEC2015** [24].

Visual Perception

The integrated conscious response and interpretation of the total visual stimulus situation. This response may be modified or interpreted in terms of stored physiological and psychological remnants of past experience which are brought to bear in that situation. For more information, see **J264_OCT2021** [34].

V-MAC

Vehicle Management and Control Unit (Mack Trucks).

VMT

Vehicle miles of travel.

VOC

Volatile organic compounds.

VOQ

Vehicle Owner's Questionnaire (NHTSA).

VRDU

Video radar decision unit.

VRTC

Vehicle Research and Test Center. NHTSA's in-house laboratory for vehicle research and testing.

VSS

Vehicle speed sensor.

W

WAD

See Whiplash-Associated Disorder.

Wake-Up

See Algorithm Enable.

Wash Out

A low-side crash mode that occurs when the front wheel loses grip and slides out to one side. For more information, see **J3133_SEP2019** [18].

Waypoint

In an automated driving system (ADS), the coordinates of a physical point in space aligned to an external reference frame used for ADS navigation. For more information, see **J3131_202203** [39].

WB

See Wheelbase.

WCS

For an electric vehicle, wireless charging system.

Weight-Carrying Hitch

A mechanical and/or structural device that connects a trailer to a towing vehicle that does not employ features designed to redistribute the load imposed at the hitch and coupling connection. For more information, see **J684_MAY2014** [40].

Weight Shift

The redistribution of vertical wheel loads in response to a longitudinal or lateral accelerations acting on a vehicle due to the height of the center of mass of the vehicle. Weight shift fore/aft results from longitudinal accelerations, for example, from braking or abruptly depressing the accelerator. Weight shift side to side results from lateral acceleration, for example, from vehicle cornering or side slipping.

Well

That part of the rim located with sufficient depth and width to enable the tire beads to be mounted and demounted over the mounting side rim flange. See Item 13 on Figure 19. For more information, see **J2047_201911** [15].

Wheel (Rim-Disc Assembly)

A rotating, load-carrying member between a tire and the hub, usually consisting of two major parts, the rim and the wheel disc, which may be integral, permanently attached, or detachable. For more information, see **J2047_201911** [15].

Wheel (Tire-Wheel Assembly)

An assembly consisting of the wheel disc, rim, an inflated tire, nuts, bolts, valve, and balance weights capable of (a) rotating about an axle; (b) carrying the load supported by the axle; (c) generating tire shear (tractive) forces between the tire tread surface and the road surface, necessary for control of vehicle motion; and (d) transmitting disturbance forces due to road surface irregularities. For more information, see **J2047_201911** [15].

Wheelbase (WB)

For a light vehicle, the perpendicular distance between the center of the front axle to the center of the rear axle; for a heavy truck or truck tractor with tandem axles, the distance between the center of the front axle and the midpoint between the centerlines of the tandem axles. In both cases, the measurements are taken with zero steer at steerable axle(s). See Figure 3.

Wheelbase, Trailer

The distance between the vertical projections of the articulation point and the center of the trailer axle onto the ground plane, with the tow-vehicle/trailer combination at rest on a horizontal surface, at a prescribed load condition, set of vehicle trim heights, or set of suspension trim heights. For more information, see **J670_JUN2022** [6].

Wheel Contact Center (Center-of-Tire Contact, Tire Coordinate System Origin)

This is the tire coordinate system origin. It is defined by the geometry of the wheel mounted on the spindle and is the intersection of the contact line and the normal projection of the wheel-spin axis onto the road plane. For more information, see **J2047_201911** [15].

Wheel Contact Line

The intersection of the wheel plane and the road plane. For more information, see **J2047_201911** [15].

Wheel Hop

The vertical oscillatory motion of a wheel between the road surface and the sprung mass. For more information, see **J670_JUN2022** [6].

Also, the tendency of a tire and wheel assembly to hop during braking or acceleration of a vehicle. For more information, see *Motor Truck Engineering Handbook* [7].

Wheel Loaded Radius (Wheel Center Height), R_L

The distance between wheel center and the center of contact in the wheel plane at a specified operating condition (load, speed, etc.). R_L is reported as an absolute value (Figure 48). For more information, see **J2047_201911** [15].

Note: On a flat road surface, loaded radius is equal to the moment arm of the shear force or its longitudinal or lateral components with respect to the wheel center. However, on a curved road surface, loaded radius is not equal to the moment arm of the shear force or its components. On a curved surface, the effective moment arm is ρ as shown in Figure 48.

FIGURE 48 Wheel center height and effective moment arm for a curved road surface.

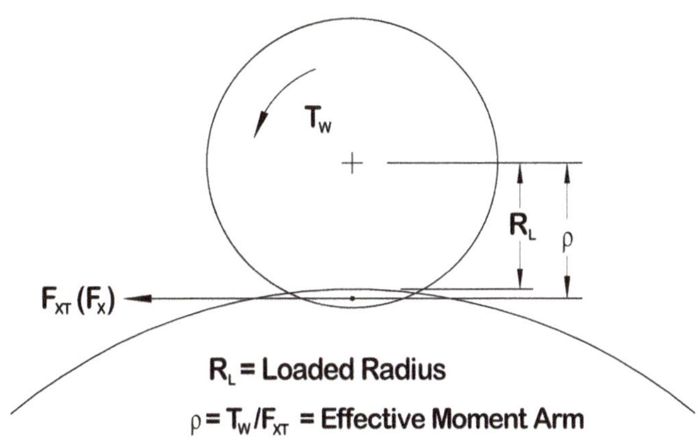

Wheel Offset

The measured distance from the attachment face of a wheel to the wheel plane. Wheel offset is positive if the mounted wheel increases the vehicle track with respect to the hub face, and it is negative if the mounted wheel decreases vehicle track with respect to the hub face. For more information, see **J2047_201911** [15].

Wheel Slip

For a vehicle traveling forward, the ratio of the forward velocity of a wheel at the road contact patch, V_{px}, to the forward velocity at the center of the wheel, V_x (for braking), or the ratio of the forward velocity of a wheel at the contact patch, V_{px}, to the difference between the velocity of the wheel at the contact patch and the forward velocity at the center of the wheel, $V_x - V_{px}$ (for traction or acceleration). Thus, slip for a braking wheel is $s_b = V_{px} / V_x$ and slip for an accelerating wheel is $s_a = V_{px} / (V_x - V_{px})$. Parameters are shown in Figure 49.

The two alternative definitions of wheel slip presented here maintain slip numerically, and intuitively, in both cases, between 0 (no slip) and 1 (maximum slip). Other definitions of wheel slip exist but can have it numerically vary between 0 (no slip) and ∞ (maximum slip).

FIGURE 49 Diagram showing the various parameters associated with wheel kinematics and wheel slip.

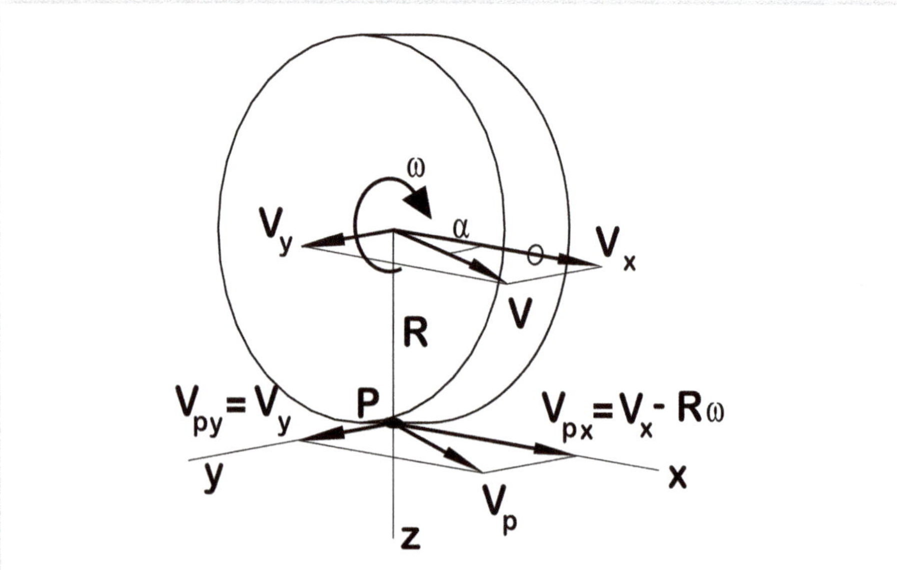

Wheel Slip Angle (Also Wheel Sideslip Angle)

The angle between a wheel's heading axis (x-axis) and direction of the velocity vector of the center of the wheel.

Wheel Slip Coefficient

The slope of the initial linear portion of the longitudinal force-wheel slip curve of a tire.

Wheel Speed Sensor

A device to sense the rotational speed of a wheel and transmit that signal to the electronic controller. The most common sensors currently in use are magnetic pickup sensors positioned near a wheel that produce a signal to indicate wheel speed to the electronic control unit (ECU). In general, a sensor generates an electrical signal with a frequency proportional to the wheel speed. A permanent magnet and passing metal teeth combine to produce a signal. The teeth alter the magnetic field produced by the sensor. In antilock brake system (ABS) applications, these signals are relayed to the ECU, which processes them to determine wheel speed. For more information, see *Commercial Vehicle Braking Systems: Air Brakes, ABS and Beyond* [5].

Wheel Track

Lateral distance between tire centers. See Track Width. For more information, see *Motor Truck Engineering Handbook* [7].

Wheel Wobble

A self-excited oscillation of a pair of steerable wheels about their steering axes, occurring without appreciable tramp. For more information, see **J670_JUN2022** [6].

Whiplash-Associated Disorder (WAD)

Caused by an acceleration-deceleration mechanism of energy transfer to the neck; may stem from a motor vehicle collision, as the impact may cause bony or soft tissue injuries (whiplash injury) in occupants that lead to a variety of clinical manifestations. For more information, see *Motor Vehicle Collision Injuries: Biomechanics, Diagnosis, and Management* [20].

Windshield Header

The structural body member that connects the upper portions of the left and right A pillars and is above the top edge of the windshield.

Wobble Mode

On a motorcycle, primarily, a motion of the front caster assembly about the steer axis. It can be lightly damped with a natural frequency in the range of 6 to 10 Hz, depending on front caster assembly properties and other vehicle characteristics. The damping can vary significantly with speed and front caster assembly properties and may become oscillatory or unstable under unusual conditions. For more information, see **J3133_SEP2019** [18].

Work-Energy Principle

This principle states that the work done by the sum of all the forces acting on a body equals the change in the kinetic energy of the body. This principle is used to derive the formula $v_f = \sqrt{v_i - 2fgd}$ used frequently in the field of vehicle crash reconstruction.

World Model

An automated driving system's internal representation of the portions of the environment of which the system is aware, or that area of interest to the system and/or the user for the purpose of achieving a specific goal. For more information, see **J3131_202203** [39].

WOT

Wide-open throttle. Also known as full throttle, the fully opened condition of the throttle on an internal combustion engine.

Wrap Pedestrian Collision

A frontal collision of a vehicle and pedestrian or a vehicle and cyclist in which the initial contact occurs at a point below the center of gravity of the pedestrian or cyclist and where the frontal geometry of the vehicle allows the pedestrian to move rearward relative to the vehicle and strike another portion of the vehicle, such as a windshield. The latter impact causes the pedestrian or cyclist to develop an airborne trajectory, followed by an impact with the ground and tumbling and sliding to rest.

WWC
Windshield wiper control.

WWDW
Wrong-way driving warning.

X

XBR
External brake request.

X Direction
In the direction of a vehicle's X axis, which is parallel to the vehicle's longitudinal centerline. The X direction is positive in the direction of forward vehicle travel. See Y Direction.

Y

Yaw

See Pitch, Roll, Yaw; Yaw Angle.

Yaw Angle

The angle between the heading of a vehicle and a fixed reference. See Figure 24.

Yaw Mark

A tire mark caused by a sideslipping tire, often showing a striped pattern called striations.

Yaw Moment of Inertia

The moment of inertia about a vertical axis of a vehicle. See Moment of Inertia; Radius of Gyration.

Yaw Rate

Angular velocity about the z-axis. See Figure 24.

Y Direction

In the direction of a vehicle's Y axis, which is perpendicular to its X axis and in the same horizontal plane as that axis. The Y direction is positive from left to right, from the perspective of the driver when seated in the vehicle facing the direction of forward vehicle travel. See X Direction.

Z

Z Direction

In the direction of a vehicle's Z axis, which is perpendicular to the X and Y axes. The Z direction is positive in a downward direction.

References

1. Brach, R. Matthew, Raymond M. Brach, and James J. Mason, *Vehicle Accident Analysis and Reconstruction Methods*, 3rd ed. (Warrendale: SAE International, R-516, 2022).
2. Society of Automotive Engineers, Standard J2939_201304, "Child Passenger Safety Glossary," 2013.
3. Nordin, Margareta, and Victor H. Frankel, *Basic Biomechanics of the Musculoskeletal System* (Philadelphia: Lea and Febiger, 1989).
4. Society of Automotive Engineers, Standard J885_FEB2011, "Human Tolerance to Impact Conditions as Related to Motor Vehicle Design," 2013.
5. Buckman, Leonard C., *Commercial Vehicle Braking Systems: Air Brakes, ABS and Beyond* (Warrendale: SAE International, SP-1405, 1998), ISBN of 0-7680-0330-X.
6. Society of Automotive Engineers, Standard J670_JUN2022, "Vehicle Dynamics Terminology," 2022.
7. Fitch, James William, *Motor Truck Engineering Handbook*, 4th ed. (Warrendale: SAE International, R-125, 1994), ISBN 1-56091-378-9.
8. Dukkipati, Rao, et Al., *Road Vehicle Dynamics* (Warrendale: SAE International, R-366, 2008), ISBN 978-0-7680-1643-7.
9. Society of Automotive Engineers, Standard J3016_JUN2018, "Taxonomy and Definitions for Terms Related to Driving Automation Systems for On-Road Motor Vehicles," 2018.
10. Society of Automotive Engineers, Standard J2399_OCT2021, "Adaptive Cruise Control (ACC) Operating Characteristics and User Interface," 2021.
11. Society of Automotive Engineers, Standard J3216_JUL2021, "Taxonomy and Definitions for Terms Related to Cooperative Driving Automation for On-Road Motor Vehicles," 2021.
12. 49 Code of Federal Regulations §393.5 Definitions.

13. Society of Automotive Engineers, Standard J826_JUN2021, "Devices for Use in Defining and Measuring Vehicle Seating Accommodation," 2021.

14. Society of Automotive Engineers, Standard J1100_NOV2009, "Motor Vehicle Dimensions," 2021.

15. Society of Automotive Engineers, Standard J2047_NOV2019, "Tire Performance Terminology," 2019.

16. Society of Automotive Engineers, Standard J1803_JAN2013, "Seat Belt Restraint System Hardware—Glossary of Terms," 2013.

17. Society of Automotive Engineers, Standard J1715_SEP2022, "Hybrid Electric Vehicle (HEV) and Electric Vehicle (EV) Terminology," 2022.

18. Society of Automotive Engineers, Standard J3133_SEP2019, "Motorcycle Terminology," 2019.

19. Chaffin, Don B., Gunnar B. J. Anderson, and Bernard J. Martin, *Occupational Biomechanics*, 3rd ed. (Hoboken, NJ: Wiley, 1999).

20. Nordhoff, Larry S., *Motor Vehicle Collision Injuries: Biomechanics, Diagnosis, and Management*, 2nd ed. (Sudbury, MA: Jones and Bartlett, 2005).

21. Society of Automotive Engineers, Standard J2990_NOV2020, "Hybrid and Electric Vehicle Safety Systems Information Report," 2020.

22. Society of Automotive Engineers, Standard J3157_FEB2019, "Active Safety Bicyclist Test Targets Recommendation," 2019.

23. Society of Automotive Engineers, Standard J3063_MAR2021, "Active Safety Systems Terms and Definitions," 2021.

24. Society of Automotive Engineers, Standard J3077_DEC2015, "Definitions and Data Sources for the Driver Vehicle Interface (DVI)," 2015.

25. Huston, Ronald, *Principles of Biomechanics* (Boca Raton, FL: Taylor and Francis, 2009).

26. Giapponi, Thomas, *Tire Forensic Investigation* (Warrendale: SAE International, R-387, 2008), ISBN 978-0-7680-1955-1.

27. Society of Automotive Engineers, Standard J3194_NOV2019, "Taxonomy and Classification of Powered Micromobility Vehicles," 2019.

28. Society of Automotive Engineers, Standard J2944_FEB2023, "Operational Definitions of Driving Performance Measures and Statistics," 2023.

29. Baker, J. Stannard, "Glossary," *Traffic Accident Investigation Manual* (Evanston, IL: Northwestern University Traffic Institute, 1975), pp. 313–321.

30. Society of Automotive Engineers, Standard J2578_AUG2014, "Recommended Practice for General Fuel Cell Vehicle Safety," 2014.

31. Society of Automotive Engineers, Standard J224_MAY2022, "Collision Deformation Classification," 2022.
32. Society of Automotive Engineers, Standard J3114_DEC2016, "Human Factors Definitions for Automated Driving and Related Research Topics," 2016.
33. Avallone, Eugene A., and Theodore Baumeister III, *Marks' Standard Handbook for Mechanical Engineers*, 10th ed. (New York: McGraw-Hill, 1996), pp. 6–163.
34. Society of Automotive Engineers, Standard J264_OCT2021, "Vision Glossary," 2021.
35. Society of Automotive Engineers, Standard J2969_201701, "Use of the Critical Speed Formula," 2017.
36. Society of Automotive Engineers, Standard J2926_201707, "Rollover Testing Methods," 2017.
37. *Bosch Automotive Handbook*, 9th ed. (www.bentleypublishers.com, 2014), p. 758.
38. Society of Automotive Engineers, Standard J3198_202010, "Driver Drowsiness and Fatigue in the Safe Operation of Vehicles—Definition of Terms and Concepts," 2020.
39. Society of Automotive Engineers, Standard J3131_202203, "Definitions for Terms Related to Automated Driving Systems Reference Architecture," 2022.
40. Society of Automotive Engineers, Standard J684_MAY2014, "Trailer Couplings, Hitches, and Safety Chains—Automotive Type," 2014.
41. Society of Automotive Engineers, Standard J1698_MAR2023, "Event Data Recorder," 2023.
42. 49 Code of Federal Regulations §565.12 Definitions.
43. 49 Code of Federal Regulations §571.3 Definitions.
44. 49 Code of Federal Regulations §390.5 Definitions.
45. 49 Code of Federal Regulations §523.2 Definitions.
46. 49 Code of Federal Regulations §563.5 Definitions.
47. Society of Automotive Engineers, Standard J2728_NOV2020, "Heavy Vehicle Event Data Recorder (HVEDR)," 2020.
48. Dowling, Norman, *Mechanical Behavior of Materials* (Hoboken, NJ: Prentice-Hall, 1993), p. 339.
49. Wulpi, Donald J., *Understanding How Components Fail* (Materials Park, OH: ASM International, 1985), p. 160.

50. Shigley, Joseph E., and Charles R. Mischke, *Standard Handbook of Machine Design*, 2nd ed. (New York: McGraw-Hill, 1996), p. 107.

51. *Quick Reference Guide (2010 Version) to Federal Motor Vehicle Safety Standards and Regulations*, DOT HS 811 439 (Washington, DC: NHTSA, 2011), p. ii.

52. *ANSI D16.1 Manual on Classification of Motor Vehicle Traffic Accidents*, 6th ed. (Chicago: National Safety Council, 2017).

53. *Potential Failure Mode and Effects Analysis (FMEA) Reference Manual*, 2nd ed. (Southfield, MI: AIAG, 1995), p. 1.

54. Society of Automotive Engineers, Standard J1626_MAR2023, "Recommended Practice for Braking, Stability, and Control Performance Test Procedure for Air Brake-Equipped Truck-Tractors," 2023.

55. Society of Automotive Engineers, Standard J2358_AUG2022, "Low-Speed Vehicles," 2022.

56. *Bosch Automotive Handbook*, 9th ed. (www.bentleypublishers.com, 2014), p. 475.

57. Society of Automotive Engineers, Standard J2258_JUL2022, "Light Utility Vehicles," 2022.

58. Society of Automotive Engineers, Standard J2344_OCT2020, "Guidelines for Electric Vehicle Safety," 2020.

59. Avallone, Eugene A., and Theodore Baumeister III, *Marks' Standard Handbook for Mechanical Engineers*, 10th ed. (New York: McGraw-Hill, 1996), pp. 7–12.

60. Hershman, Lawrence L., "The U.S. New Car Assessment Program (NCAP): Past, Present and Future," National Highway Traffic Safety Administration, Paper Number 390, 2001.

61. Society of Automotive Engineers, Best Practice AVSC00006202103, "AVSC Best Practice for Metrics and Methods for Assessing Safety Performance of Automated Driving Systems (ADS)," 2021.

62. *Air Brake Handbook*, Bendix Commercial Vehicle Systems, www.bendix.com, 2009, p. 21.

63. Society of Automotive Engineers, Standard J2188_201807, "Commercial Truck and Bus SAE Recommended Procedure for Vehicle Performance Prediction and Charting," 2018.

64. Chan, Ching-Yao, *Fundamentals of Crash Sensing in Automotive Air Bag Systems* (Warrendale: SAE International, R-217, 2000), ISBN 0-7680-0499-3.

65. Society of Automotive Engineers, Standard J3151_201810, "Relating Experimental Drive Distraction and Driving Performance Metrics to Crash Involvement—Definitions of Terms and Concepts," 2018.

66. Society of Automotive Engineers, Standard J2388_202208, "Secondary Control Modifications," 2022.

67. Society of Automotive Engineers, Standard J383_202207, "Motor Vehicle Seat Belt Anchorages—Design Recommendations," 2022.

68. Society of Automotive Engineers, Standard J1301_202206, "Truck Deformation Classification," 2022.

Appendix A: Units and Numbers

Use of SI (Metric Units of Measure in SAE Technical Papers)

The long-term goal for SAE is international communication with minimal effort and confusion. Therefore, the use of SI units in all technical publications and presentations is preferred. The Society will strive toward universal usage of SI units and will encourage their use whenever appropriate.

However, the Society also recognizes that sectors of the mobility market do not yet use SI units because of tradition, regulatory language, or other reasons. Mandating the use of SI units in these cases will impede rather than facilitate technical communication. Therefore, it is the policy to allow non-SI units and dual dimensioning where communication will be enhanced. This shall not be viewed as an avenue to circumvent the long-term goal of 100% SI usage.

Instructions on SAE-approved techniques for conversion of units are contained in "SAE Recommended Practices, Rules for SAE Use of SI (METRIC) Units—TSB003." Copies of TSB003 can be obtained from SAE headquarters (Table A.1).

Although what follows represents a change to the current policy, it is not a change to the SAE Board of Directors' Policy since it falls within the scope of the words, "where a conflicting industry practice exists." Dual (metric/US customary) units for the following vehicle characteristics may be considered where communication will be enhanced.

TABLE A.1 Metric and US customary units.

Vehicle characteristic	Metric units	US customary units
Volume, engine displacement	Liters, L, or cubic cm, cm^3	Cubic inches, $in.^3$
Liquid volume	Liters, L	Pints/quarts/gallons
Engine power	Kilowatts, kW	Brake horsepower, bhp
Engine torque	Newton meters, N - m	Foot-pounds, lb-ft
Mass	Kilograms, kg	Slugs, lb-s^2/ft
Pressure, stress	Kilopascal, kPa	Pounds per square inch, psi
Temperature	Degrees Celsius, °C	Degrees Fahrenheit, °F
Area	Square cm, cm^2	Square inches, $in.^2$
Linear dimensions	Millimeters, mm, meters, m, or kilometers, km	Inches, in., feet, ft, miles, mi
Spring rates	Newtons per millimeter, N/mm	Pounds per inch, lb/in.
Speed	Kilometers per hour, km/h or kph	Miles per hour, mph
Fuel economy	Kilometers per liter, km/L or kmpL	Miles per gallon, mpg
Force	Newtons, N	Pounds, lb
Acceleration	Kilometers per second per second, km/s^2, g	Feet per second per second, ft/s^2, g

Numbers, Significant Figures, and Rounding

Significant Figures

In all branches of science and technology, numbers are used to express values, that is, levels or amounts of physical quantities. It is important to state numbers appropriately so that they properly convey the intended information. The number of significant figures contained in a stated number reflects the accuracy to which that quantity is known. For example, suppose the speed of a vehicle is reported as 21 m/s (69 ft/s). Is 21 m/s different from 21.0 m/s? According to the rules of significant figures, yes, but in practice, it may or may not. Could the number 21 m/s imply 20.9 m/s or less, or could it imply 21.1 m/s or greater? It could, but such implications or interpretations must be determined from context, not the number 21 itself. Answers to some of these questions are related to the topic of uncertainty (covered in Chapter 1). To properly quantify and communicate a physical measurement or property, it should be stated as a reference value plus and minus an uncertainty. For example, a speed stated as $v = 21.0 \pm 0.6$ m/s clearly is meant to be between 20.4 m/s and 21.6 m/s. This is one of the ways of estimating and revealing the uncertainty of results. But the basic rules of using significant figures and rounding must be understood before uncertainty can be expressed. Some of the rules for handling and interpreting the significance of numbers are covered in this Appendix. Note that the terms *significant figures* and *significant digits* are used synonymously.

The number of significant figures in a number is defined in the following way [A.1, A.2]:

1. The leftmost nonzero digit of a number is the most significant digit.
2. If there is no decimal point, the rightmost nonzero digit is the least significant digit.
3. If there is a decimal point, the rightmost digit is the least significant digit, even if it is a zero.
4. All digits, from the least to the most significant, are counted as significant.

So, for example, 2.610. and 2,498 have four significant digits each, whereas 0.125 and 728,000 have three significant digits. The following numbers each have five significant digits: 1000.0, 1206.5, 12,065,000, and 0.00012065. Unless it is stated to be exact, the speed of 21 m/s has two significant figures. If it is *exact*, then 21 is equivalent to 21.0000 ..., with an unlimited number of zeros. Each of the speeds 20.4 and 21.6 has three significant figures.

When numbers are very large or very small, it is convenient to express them in *scientific notation*. To use scientific notation, a decimal point is placed immediately after the leftmost significant digit and the number is given a suffix of 10 raised to a power n. The value of n is positive or negative. If the magnitude (disregarding the sign) of the stated number is less than 1, then $n < 0$. If the stated number is greater than 10, $n > 0$. If the stated number is between 1 and 10, $n = 0$. The value of n is the power of 10 that returns the number in scientific notation to its original value. For example, 0.0000687 becomes 6.87×10^{-5} and 12,360,000 becomes 1.236×10^{7}. Note that the number of significant digits does not change when converting to or from scientific notation.

Rounding of Numbers

After completing calculations or when listing the results of measurements, it usually is necessary to round numbers to a lesser number of significant figures by discarding digits. Three possibilities can arise; these are:

1. *The leftmost discarded digit is less than 5.* When rounding such numbers, the last digit retained should remain unchanged. For example, if 3.46325 is to be rounded to four digits, the digits 2 and 5 would be discarded and 3.463 remains.
2. *The leftmost discarded digit is greater than 5 or it is a 5 followed by at least one digit other than 0.* In such cases, the last figure retained should be increased by one. For example, if rounded to four digits, 8.37652 would become 8.377; if rounded to three digits, it would be 8.38.

3. *The leftmost discarded digit is a 5, followed only by zeros or no other numbers.* Here, the last digit retained should be rounded up if it is an odd number, but no adjustment made if it is an even number. For example, 21.165, when rounded to four significant digits, becomes 21.16. The number 21.155 would likewise round to the same value, 21.16.

A reason for this last rule [A.2] is to avoid systematic errors that otherwise would be introduced into the average of a group of such numbers. Not all computer software follows this rule, however,[1] and when rounding for purposes of reporting results of measurements and/or calculations, the even-odd rule is not critical.

Consistency of Significant Figures When Adding and Subtracting

When adding and subtracting numbers, proper determination of the number of significant figures is stated as a rule [A.1]. The rule is *the answer shall contain no significant digits farther to the right than occurs in the number with the least significant digits.* The simplest way of following this rule is first to add or subtract the numbers using all of the stated significant figures[2] followed by rounding of the final answer. For example, consider the addition of the three numbers, 964,532, 317,880, and 563,000. These have six, five, and three significant figures, respectively. The sum by direct addition is 1,845,412. The answer then is adjusted, or rounded, to conform to the number with the least significant figures (563,000 with three), giving the final result, 1,845,000. This number has no more zero digits to the right of the comma than does 563,000. Now consider the sum of the three numbers, 964,532, –317,880, and –563,000; the direct result is 83,652. As above, this must be made to conform with the significant figures of 563,000 by using the rounding rule and is 84,000.

In the last example, the concept being conveyed is that the number 563,000 is "indefinite" to the right of the "3" digit. It is not known if 563,000 could really mean 562,684 or 563,121 or other values because 563,000, itself, may have been obtained by rounding. If it had been stated as 563,000.0, then everything would be different (since 563,000.0 would have seven significant figures and 317,880 would then have the least significant digits of the three numbers to be added in the above example).

---

[1] The reader may wish to try such an example in their favorite software.

[2] ASTM SI-10 suggests first rounding each individual number to one significant figure greater than the least *before* adding or subtracting and then rounding the final answer. Though this may be better, it is not the way most computer software operates. Rounding after summing typically gives the same result.

Consistency of Significant Figures When Multiplying and Dividing

ASTM SI-10 [A.1] states a rule for multiplying and dividing as the product or quotient shall contain no more significant digits than are contained in the number with the fewest significant digits. For example, consider the product, $125.64 \times 829.4 \times 1.25$, of the three numbers with five, four, and three significant digits, respectively. The answer from straightforward multiplication is 130,257.27. After rounding to three significant figures, the proper end result of the multiplication is 130,000. Note that the answer, 130,000, by itself appears to have only two significant figures. This illustrates that ambiguities sometimes can arise when determining significant figures and that the amount of significant figures of a number may need to be found from context. A way of resolving such ambiguities is to express results of rounding in scientific notation. In this case the result would be 1.30×10^5.

Other Forms of Number Manipulation

Not all calculations are done with addition, subtraction, multiplication, and division. There is the taking of roots, logarithms, trigonometric functions, etc. In addition, sometimes strict adherence to rounding rules can produce paradoxical or impractical results (see the following example). So more general rules are needed. In summary, two very general but some *practical* rules are recommended:

1. In rounding of numbers and conversion of units, retain a number of significant digits such that accuracy and precision are neither sacrificed nor exaggerated.
2. When making and reporting calculations, continually carry all of the significant figures of a calculating device without rounding intermediate values, and round only the final answer.
3. Unit conversion should precede rounding.
4. Whenever possible, explicitly state the uncertainty of the results of measurements and calculations.

Example A.1

Suppose a vehicle skids to a stop over a distance of d = 33.9 m from an initial speed, v, on a pavement with a uniform frictional drag coefficient of $f = 0.7 \pm 0.1$. Use the minimum and maximum values of f and Equation (1.1) to calculate bounds on the initial speed. Convert the results to US customary units of feet per second (ft/s).

Solution The lower value of speed for $f = 0.6$ is

$$v = \sqrt{2fgd} = \sqrt{2 \times 06 \times 9.806650 \times 33.9} = 19.973345\ldots$$

Similarly, the initial speed for $f = 0.8$ is

$$v = \sqrt{2 \times 08 \times 9.806650 \times 33.9} = 23.063233\ldots$$

The frictional drag coefficient and its uncertainty have the fewest number of significant figures of the input values. According to the rules, the final results should be rounded to one significant figure. Rounding 19.973345... to a single significant digit gives a speed of $v = 20$ m/s. Rounding 23.063235... to a single significant digit also gives a speed of $v = 20$ m/s. Both upper and lower bounds result in the same speed, $v = 20$ m/s. Clearly, the result is an exaggeration of precision. Consider now another approach.

The variation of $f = \pm\ 0.1$ is another way of saying that because of uncertainty, f can take on any value between 0.6 and 0.8.[3] From the above discussion of significant figures and rounding, a point of view can be taken that the lower value, 0.6, for example, could be the result of rounding to one significant figure of any number from 0.55+ to 0.65− (such as 0.551, 0.642, etc.). Similarly, the upper value, 0.8, could be viewed as the result of rounding of any number from 0.75+ to 0.85− (such as 0.751, 0.842, etc.). So the full range of values of the frictional drag coefficient corresponding to the stated uncertainty and from the concepts of significant figures is $0.55 \le f \le 0.85$. At this point the calculations are performed *as if* all numbers are exactly giving a speed range of $19.123022\ldots \le v \le 23.773036\ldots$ m/s. Since rounding to one significant figure here produces an exaggeration of precision (as above), rounding is done to an additional significant figure. Consequently, the final result is stated as $19 \le v \le 24$ m/s, or $v = 19.5 \pm 2.5$ m/s. Precision no longer is exaggerated. An initial ±14% variation (0.7 ± 0.1) becomes a 12% variation of v(19.5 ± 2.5) through the use of Equation (1.1).

Finally, the speed is to be converted to the US customary units of feet per second (ft/s). The proper conversion factor is 1 ft = 0.3048 m (this is an *exact* conversion; see the following unit conversion table). Unit conversions should be done before rounding, so $19.123022\ldots \le v \le 23.773036\ldots$ m/s becomes $62.739573\ldots \le v \le 77.995525\ldots$ ft/s. Rounding again to one significant figure gives the same result, 70 ft, so another significant figure is acceptable, giving $63 \le v \le 78$ ft/s, or $v = 70.5 \pm 7.5$ ft/s.

[3] Note that there is no implication of the likelihood of any of the values within this range.

Another consideration that must be kept in mind when rounding is the use or purpose of the results, for example, if the speed calculated in the last example is to be compared to a speed limit, say, 25 m/s. Rounding to a number of significant digits to the right of the decimal point is superfluous. The result $19 \leq v \leq 24$ m/s is satisfactory to conclude that the calculated speed is less than the speed limit. Instead, suppose that the calculated speed is a measure of vehicle braking performance and is to be compared to a governmental regulation stated to three significant figures. Rounding to an additional significant figure leads to an exaggeration of accuracy. To compare the speed to such a regulation requires a more accurate value of friction, stated at least to two significant figures.

Unit Conversions for Common Units

Factors in boldface are exact. When options exist, units in the first column printed in *italics* are preferred by the National Institute for Standards and Technology [A.3].

To convert from	To	Multiply by	
Acre (based on US survey foot)	Square meter (m^2)	4.046 873	E+03
Acre foot (based on US survey foot)	Cubic meter (m^3)	1.233 489	E+03
Ampere hour (A·h)	Coulomb (C)	**3.6**	**E+03**
Atmosphere, standard (atm)	Pascal (Pa)	**1.013 25**	**E+05**
Atmosphere, standard (atm)	Kilopascal (kPa)	**1.013 25**	**E+02**
Atmosphere, technical (at)	Pascal (Pa)	**9.806 65**	**E+04**
Atmosphere, technical (at)	Kilopascal (kPa)	**9.806 65**	**E+01**
Bar (bar)	Pascal (Pa)	**1.0**	**E+05**
Bar (bar)	Kilopascal (kPa)	**1.0**	**E+02**
Barn (b)	Square meter (m^2)	**1.0**	**E−28**
Barrel [for petroleum, 42 gallons (US)] (bbl)	Cubic meter (m^3)	1.589 873	E−01
Barrel [for petroleum, 42 gallons (US)] (bbl)	Liter (L)	1.589 873	E+02
British thermal unit (mean) (Btu)	Joule (J)	1.055 87	E+03
Bushel (US) (bu)	Cubic meter (m^3)	3.523 907	E−02
Bushel (US) (bu)	Liter (L)	3.523 907	E+01
Calorie (cal) (mean)	Joule (J)	4.190 02	E+00
Candela per square inch (cd/in.^{2})	Candela per square meter (cd/m^2)	1.550 003	E+03
Carat, metric	Kilogram (kg)	**2.0**	**E−04**
Carat, metric	Gram (g)	**2.0**	**E−01**

To convert from	To	Multiply by	
Centimeter of mercury (0°C)	Pascal (Pa)	1.333 22	E+03
Centimeter of water (4°C)	Pascal (Pa)	9.806 38	E+01
Centimeter of water, conventional (cm H_2O)	Pascal (Pa)	**9.806 65**	**E+01**
Centipoise (cP)	Pascal second (Pa·s)	**1.0**	**E−03**
Centistokes (cSt)	Meter squared per second (m^2/s)	**1.0**	**E−06**
Chain (based on US survey foot) (ch)	Meter (m)	2.011 684	E+01
Circular mil	Square meter (m^2)	5.067 075	E−10
Cord (128 ft^3)	Cubic meter (m^3)	3.624 556	E+00
Cubic foot (ft^3)	Cubic meter (m^3)	2.831 685	E−02
Cubic inch ($in.^3$)	Cubic meter (m^3)	1.638 706	E−05
Cubic mile (mi^3)	Cubic meter (m^3)	4.168 182	E+09
Cubic yard (yd^3)	Cubic meter (m^3)	7.645 549	E−01
Cup (US)	Cubic meter (m^3)	2.365 882	E−04
Cup (US)	Liter (L)	2.365 882	E−01
Day (d)	Second (s)	**8.64**	**E+04**
Day (sidereal)	Second (s)	8.616 409	E+04
Degree (angle) (°)	Radian (rad)	1.745 329	E−02
Degree Celsius (temperature) (°C)	Kelvin (K)	K = °C + **273.15**	
Degree Celsius (temperature interval) (°C)	Kelvin (K)	**1.0**	**E+00**
Degree centigrade (temperature)	Degree Celsius (°C)	°C = deg. cent.	
Degree centigrade (temperature interval)	Degree Celsius (°C)	1.0	E+00
Degree Fahrenheit (temperature) (°F)	Degree Celsius (°C)	°C = (°F − **32**)/**1.8**	
Degree Fahrenheit (temperature) (°F)	Kelvin (K)	K = (°F + **459.67**)/**1.8**	
Degree Fahrenheit (temperature interval) (°F)	Degree Celsius (°C)	5.555 556	E−01
Degree Fahrenheit (temperature interval) (°F)	Kelvin (K)	5.555 556	E−01
Degree Rankine (°R)	Kelvin (K)	K = (°R)/**1.8**	
Degree Rankine (temperature interval) (°R)	Kelvin (K)	5.555 556	E−01
Denier	Kilogram per meter (kg/m)	1.111 111	E−07
Dyne (dyn)	Newton (N)	**1.0**	**E−05**
Dyne centimeter (dyn·cm)	Newton meter (N·m)	**1.0**	**E−07**

To convert from	To	Multiply by	
Dyne per square centimeter (dyn/cm^2)	Pascal (Pa)	**1.0**	**E-01**
Erg (erg)	Joule (J)	**1.0**	**E-07**
Erg per second (erg/s)	Watt (W)	**1.0**	**E-07**
Fathom (based on US survey foot)	Meter (m)	1.828 804	E+00
Fluid ounce (US) (fl oz)	Cubic meter (m^3)	2.957 353	E-05
Fluid ounce (US) (fl oz)	Milliliter (mL)	2.957 353	E+01
Foot (ft)	Meter (m)	**3.048**	**E-01**
Foot (US survey) (ft)	Meter (m)	3.048 006	E-01
Footcandle	Lux (lx)	1.076 391	E+01
Footlambert	Candela per square meter (cd/m^2)	3.426 259	E+00
Foot of water, conventional (ftH$_2$O)	Pascal (Pa)	2.989 067	E+03
Foot of water, conventional (ftH$_2$O)	Kilopascal (kPa)	2.989 067	E+00
Foot per hour (ft/h)	Meter per second (m/s)	8.466 667	E-05
Foot per minute (ft/min)	Meter per second (m/s)	**5.08**	**E-03**
Foot per second (ft/s)	Meter per second (m/s)	**3.048**	**E-01**
Foot per second squared (ft/s^2)[4]	Meter per second squared (m/s^2)^2	**3.048**	**E-01**
Foot-poundal	Joule (J)	4.214 011	E-02
Foot pound-force (ft·lbf)	Joule (J)	1.355 818	E+00
Foot pound-force per hour (ft·lbf/h)	Watt (W)	3.766 161	E-04
Foot pound-force per minute (ft·lbf/min)	Watt (W)	2.259 697	E-02
Foot pound-force per second (ft·lbf/s)	Watt (W)	1.355 818	E+00
Gal (Gal)	Meter per second squared (m/s^2)	**1.0**	**E-02**
Gallon [Canadian and UK (Imperial)] (gal)	Cubic meter (m^3)	**4.546 09**	**E-03**
Gallon [Canadian and UK (Imperial)] (gal)	Liter (L)	**4.546 09**	**E+00**
Gallon (US) (gal)	Cubic meter (m^3)	3.785 412	E-03
Gallon (US) (gal)	Liter (L)	3.785 412	E+00
Gallon (US) per day (gal/d)	Cubic meter per second (m^3/s)	4.381 264	E-08

[4] Standard value of free-fall acceleration is $g = 9.80665$ m/s^2.

To convert from	To	Multiply by	
Gallon (US) per day (gal/d)	Liter per second (L/s)	4.381 264	E-05
Gallon (US) per horsepower-hour [gal/(hp·h)]	Cubic meter per joule (m^3/J)	1.410 089	E-09
Gallon (US) per horsepower-hour [gal/(hp·h)]	Liter per joule (L/J)	1.410 089	E-06
Gallon (US) per minute (gpm) (gal/min)	Cubic meter per second (m^3/s)	6.309 020	E-05
Gallon (US) per minute (gpm) (gal/min)	Liter per second (L/s)	6.309 020	E-02
Grain (gr)	Kilogram (kg)	**6.479 891**	**E-05**
Grain (gr)	Milligram (mg)	**6.479 891**	**E+01**
Grain per gallon (US) (gr/gal)	Kilogram per cubic meter (kg/m^3)	1.711 806	E-02
Grain per gallon (US) (gr/gal)	Milligram per liter (mg/L)	1.711 806	E+01
Gram-force per square centimeter (gf/cm^2)	Pascal (Pa)	**9.806 65**	**E+01**
Gram per cubic centimeter (g/cm^3)	Kilogram per cubic meter (kg/m^3)	**1.0**	**E+03**
Hectare (ha)	Square meter (m^2)	**1.0**	**E+04**
Horsepower (550 ft·lbf/s) (hp)	Watt (W)	7.456 999	E+02
Horsepower (boiler)	Watt (W)	9.809 50	E+03
Horsepower (electric)	Watt (W)	**7.46**	**E+02**
Horsepower (metric)	Watt (W)	7.354 988	E+02
Horsepower (UK)	Watt (W)	7.4570	E+02
Horsepower (water)	Watt (W)	7.460 43	E+02
Hour (h)	Second (s)	**3.6**	**E+03**
Hour (sidereal)	Second (s)	3.590 170	E+03
Hundredweight (long, 112 lb)	Kilogram (kg)	5.080 235	E+01
Hundredweight (short, 100 lb)	Kilogram (kg)	4.535 924	E+01
Inch (in.)	Meter (m)	**2.54**	**E-02**
Inch (in.)	Centimeter (cm)	**2.54**	**E+00**
Inch of mercury, conventional (in. Hg)	Pascal (Pa)	3.386 389	E+03
Inch of mercury, conventional (in. Hg)	Kilopascal (kPa)	3.386 389	E+00
Inch of water, conventional (in. H_2O)	Pascal (Pa)	2.490 889	E+02
Kelvin (K)	Degree Celsius (°C)	t/°C = T/K - **273.15**	
Kilocalorie (mean) (kcal)	Joule (J)	4.190 02	E+03
Kilogram-force (kgf)	Newton (N)	**9.806 65**	**E+00**

To convert from	To	Multiply by	
Kilogram-force meter (kgf·m)	Newton meter (N·m)	**9.806 65**	**E+00**
Kilogram-force per square centimeter (kgf/cm^2)	Kilopascal (kPa)	**9.806 65**	**E+01**
Kilogram-force per square meter (kgf/m^2)	Pascal (Pa)	**9.806 65**	**E+00**
Kilometer per hour (km/h)	Meter per second (m/s)	2.777 778	E-01
Kilopond (kilogram-force) (kp)	Newton (N)	**9.806 65**	**E+00**
Kilowatt hour (kW·h)	Joule (J)	**3.6**	**E+06**
Kilowatt hour (kW·h)	Megajoule (MJ)	**3.6**	**E+00**
Kip (1 kip = 1000 lbf)	Newton (N)	4.448 222	E+03
Kip (1 kip = 1000 lbf)	Kilonewton (kN)	4.448 222	E+00
Kip per square inch (ksi) (kip/in.^{2})	Pascal (Pa)	6.894 757	E+06
Kip per square inch (ksi) (kip/in.^{2})	Kilopascal (kPa)	6.894 757	E+03
Knot (nautical mile per hour)	Meter per second (m/s)	5.144 444	E-01
Lambert	Candela per square meter (cd/m^2)	3.183 099	E+03
Light-year (l.y.)	Meter (m)	9.460 73	E+15
Liter (L)	Cubic meter (m^3)	**1.0**	**E-03**
Lumen per square foot (lm/ft^2)	Lux (lx)	1.076 391	E+01
Microinch	Meter (m)	**2.54**	**E-08**
Microinch	Micrometer (μm)	**2.54**	**E-02**
Micron (μ)	Meter (m)	**1.0**	**E-06**
Micron (μ)	Micrometer (μm)	**1.0**	**E+00**
Mil (0.001 in.)	Meter (m)	**2.54**	**E-05**
Mil (0.001 in.)	Millimeter (mm)	**2.54**	**E-02**
Mil (angle)	Radian (rad)	9.817 477	E-04
Mil (angle)	Degree (°)	**5.625**	**E-02**
Mile (mi)	Meter (m)	**1.609 344**	**E+03**
Mile (mi)	Kilometer (km)	**1.609 344**	**E+00**
Mile (based on US survey foot) (mi)	Meter (m)	1.609 347	E+03
Mile (based on US survey foot) (mi)	Kilometer (km)	1.609 347	E+00
Mile, nautical	Meter (m)	**1.852**	**E+03**
Mile per gallon (US) (mpg) (mi/gal)	Meter per cubic meter (m/m^3)	4.251 437	E+05

To convert from	To	Multiply by	
Mile per gallon (US) (mpg) (mi/gal)	Kilometer per liter (km/L)	4.251 437	E−01
Mile per gallon (US) (mpg) (mi/gal)	Liter per 100 kilometers (L/100 km)	Divide 235.215 by number of miles per gallon	
Mile per hour (mi/h)	Meter per second (m/s)	**4.4704**	**E−01**
Mile per hour (mi/h)	Kilometer per hour (km/h)	**1.609 344**	**E+00**
Mile per minute (mi/min)	Meter per second (m/s)	**2.682 24**	**E+01**
Mile per second (mi/s)	Meter per second (m/s)	**1.609 344**	**E+03**
Millibar (mbar)	Pascal (Pa)	**1.0**	**E+02**
Millibar (mbar)	Kilopascal (kPa)	**1.0**	**E−01**
Millimeter of mercury, conventional (mmHg)	Pascal (Pa)	1.333 224	E+02
Millimeter of water, conventional (mm H_2O)	Pascal (Pa)	**9.806 65**	**E+00**
Minute (angle) (N)	Radian (rad)	2.908 882	E−04
Minute (min)	Second (s)	**6.0**	**E+01**
Minute (sidereal)	Second (s)	5.983 617	E+01
Ounce (avoirdupois) (oz)	Kilogram (kg)	2.834 952	E−02
Ounce (avoirdupois) (oz)	Gram (g)	2.834 952	E+01
Ounce (troy or apothecary) (oz)	Kilogram (kg)	3.110 348	E−02
Ounce (troy or apothecary) (oz)	Gram (g)	3.110 348	E+01
ounce [Canadian and UK fluid (Imperial)] (fl oz)	Cubic meter (m^3)	2.841 306	E−05
Ounce [Canadian and UK fluid (Imperial)] (fl oz)	Milliliter (mL)	2.841 306	E+01
Ounce (US fluid) (fl oz)	Cubic meter (m^3)	2.957 353	E−05
Ounce (US fluid) (fl oz)	Milliliter (mL)	2.957 353	E+01
Ounce (avoirdupois)-force (ozf)	Newton (N)	2.780 139	E−01
Ounce (avoirdupois)-force inch (ozf · in.)	Newton meter (N · m)	7.061 552	E−03
Ounce (avoirdupois)-force inch (ozf · in.)	Millinewton meter (mN · m)	7.061 552	E+00
Ounce (avoirdupois) per cubic inch (oz/in.^{3})	Kilogram per cubic meter (kg/m^3)	1.729 994	E+03
Peck (US) (pk)	Cubic meter (m^3)	8.809 768	E−03
Peck (US) (pk)	Liter (L)	8.809 768	E+00
Pennyweight (dwt)	Kilogram (kg)	1.555 174	E−03
Pennyweight (dwt)	Gram (g)	1.555 174	E+00

To convert from	To	Multiply by	
Pica (computer) (1/6 in.)	Meter (m)	4.233 333	E−03
Pica (computer) (1/6 in.)	Millimeter (mm)	4.233 333	E+00
Pica (printer's)	Meter (m)	4.217 518	E−03
Pica (printer's)	Millimeter (mm)	4.217 518	E+00
Pint (US dry) (dry pt)	Cubic meter (m^3)	5.506 105	E−04
Pint (US dry) (dry pt)	Liter (L)	5.506 105	E−01
Pint (US liquid) (liq pt)	Cubic meter (m^3)	4.731 765	E−04
Pint (US liquid) (liq pt)	Liter (L)	4.731 765	E−01
Point (computer) (1/72 in.)	Meter (m)	3.527 778	E−04
Point (computer) (1/72 in.)	Millimeter (mm)	3.527 778	E−01
Point (printer's)	Meter (m)	3.514 598	E−04
Point (printer's)	Millimeter (mm)	3.514 598	E−01
Poise (P)	Pascal second (Pa · s)	**1.0**	**E−01**
Pound (avoirdupois) (lb)	Kilogram (kg)	4.535 924	E−01
Pound (troy or apothecary) (lb)	Kilogram (kg)	3.732 417	E−01
Poundal	Newton (N)	1.382 550	E−01
Poundal per square foot	Pascal (Pa)	1.488 164	E+00
Poundal second per square foot	Pascal second (Pa · s)	1.488 164	E+00
Pound foot squared (lb · ft^2)	Kilogram meter squared (kg · m^2)	4.214 011	E−02
Pound-force (lbf)[5]	Newton (N)	4.448 222	E+00
Pound-force foot (lbf · ft)	Newton meter (N · m)	1.355 818	E+00
Pound-force foot per inch (lbf · ft/in.)	Newton meter per meter (N · m/m)	5.337 866	E+01
Pound-force inch (lbf · in.)	Newton meter (N · m)	1.129 848	E−01
Pound-force inch per inch (lbf · in./in.)	Newton meter per meter (N · m/m)	4.448 222	E+00
Pound-force per foot (lbf/ft)	Newton per meter (N/m)	1.459 390	E+01
Pound-force per inch (lbf/in.)	Newton per meter (N/m)	1.751 268	E+02
Pound-force per pound (lbf/lb) (thrust-to-mass ratio)	Newton per kilogram (N/kg)	**9.806 65**	**E+00**
Pound-force per square foot (lbf/ft^2)	Pascal (Pa)	4.788 026	E+01
Pound-force per square inch (psi) (lbf/$in.^2$)	Pascal (Pa)	6.894 757	E+03

[5] If the local value of the acceleration of free fall is taken as the standard value $g = 9.90665\ m/s^2$, then the exact conversion factor is 4.448 221 615 260 5 E + 00.

To convert from	To	Multiply by	
Pound-force per square inch (psi) (lbf/in.^{2})	Kilopascal (kPa)	6.894 757	E+00
Pound-force second per square foot (lbf·s/ft^2)	Pascal second (Pa·s)	4.788 026	E+01
Pound-force second per square inch (lbf·s/in.^{2})	Pascal second (Pa·s)	6.894 757	E+03
Pound inch squared (lb·in.^{2})	Kilogram meter squared (kg·m^2)	2.926 397	E−04
Pound per cubic foot (lb/ft^3)	Kilogram per cubic meter (kg/m^3)	1.601 846	E+01
Pound per cubic inch (lb/in.^{3})	Kilogram per cubic meter (kg/m^3)	2.767 990	E+04
Pound per cubic yard (lb/yd^3)	Kilogram per cubic meter (kg/m^3)	5.932 764	E−01
Pound per foot (lb/ft)	Kilogram per meter (kg/m)	1.488 164	E+00
Pound per foot hour [lb/(ft·h)]	Pascal second (Pa·s)	4.133 789	E−04
Pound per foot second [lb/(ft·s)]	Pascal second (Pa·s)	1.488 164	E+00
Pound per gallon [Canadian and UK (Imperial)] (lb/gal)	Kilogram per cubic meter (kg/m^3)	9.977 637	E+01
Pound per gallon [Canadian and UK (Imperial)] (lb/gal)	Kilogram per liter (kg/L)	9.977 637	E−02
Pound per gallon (US) (lb/gal)	Kilogram per cubic meter (kg/m^3)	1.198 264	E+02
Pound per gallon (US) (lb/gal)	Kilogram per liter (kg/L)	1.198 264	E−01
Pound per horsepower-hour [lb/(hp·h)]	Kilogram per joule (kg/J)	1.689 659	E−07
Psi (pound-force per square inch) (lbf/in.^{2})	Pascal (Pa)	6.894 757	E+03
Psi (pound-force per square inch) (lbf/in.^{2})	Kilopascal (kPa)	6.894 757	E+00
Quad (10^{15} Btu_{IT})	Joule (J)	1.055 056	E+18
Quart (US dry) (dry qt)	Cubic meter (m^3)	1.101 221	E−03
Quart (US dry) (dry qt)	Liter (L)	1.101 221	E+00
Quart (US liquid) (liq qt)	Cubic meter (m^3)	9.463 529	E−04
Quart (US liquid) (liq qt)	Liter (L)	9.463 529	E−01
Rad (absorbed dose) (rad)	Gray (Gy)	**1.0**	**E−02**
Revolution (r)	Radian (rad)	6.283 185	E+00
Revolution per minute (rpm) (r/min)	Radian per second (rad/s)	1.047 198	E−01
Rod (based on US survey foot) (rd)	Meter (m)	5.029 210	E+00

To convert from	To	Multiply by	
Rpm (revolution per minute) (r/min)	Radian per second (rad/s)	1.047 198	E−01
Second (angle) (°)	Radian (rad)	4.848 137	E−06
Second (sidereal)	Second (s)	9.972 696	E−01
Shake	Second (s)	**1.0**	**E−08**
Shake	Nanosecond (ns)	**1.0**	**E+01**
Slug (slug)	Kilogram (kg)	1.459 390	E+01
Slug per cubic foot (slug/ft^3)	Kilogram per cubic meter (kg/m^3)	5.153 788	E+02
Slug per foot second [slug/(ft · s)]	Pascal second (Pa · s)	4.788 026	E+01
Square foot (ft^2)	Square meter (m^2)	**9.290 304**	**E−02**
Square foot per hour (ft^2/h)	Square meter per second (m^2/s)	**2.580 64**	**E−05**
Square foot per second (ft^2/s)	Square meter per second (m^2/s)	**9.290 304**	**E−02**
Square inch (in.^{2})	Square meter (m^2)	**6.4516**	**E−04**
Square inch (in.^{2})	Square centimeter (cm^2)	**6.4516**	**E+00**
Square mile (mi^2)	Square meter (m^2)	2.589 988	E+06
Square mile (mi^2)	Square kilometer (km^2)	2.589 988	E+00
Square mile (based on US survey foot) (mi^2)	Square meter (m^2)	2.589 998	E+06
Square mile (based on US survey foot) (mi^2)	Square kilometer (km^2)	2.589 998	E+00
Square yard (yd^2)	Square meter (m^2)	8.361 274	E−01
Stokes (St)	Meter squared per second (m^2/s)	**1.0**	**E−04**
Tablespoon	Cubic meter (m^3)	1.478 676	E−05
Tablespoon	Milliliter (mL)	1.478 676	E+01
Teaspoon	Cubic meter (m^3)	4.928 922	E−06
Teaspoon	Milliliter (mL)	4.928 922	E+00
Therm (EC)	Joule (J)	**1.055 06**	**E+08**
Therm (US)	Joule (J)	**1.054 804**	**E+08**
Ton, assay (AT)	Kilogram (kg)	2.916 667	E−02
Ton, assay (AT)	Gram (g)	2.916 667	E+01
Ton-force (2000 lbf)	Newton (N)	8.896 443	E+03
Ton-force (2000 lbf)	Kilonewton (kN)	8.896 443	E+00
Ton, long (2240 lb)	Kilogram (kg)	1.016 047	E+03
Ton, long, per cubic yard	Kilogram per cubic meter (kg/m^3)	1.328 939	E+03
Ton, metric (t)	Kilogram (kg)	**1.0**	**E+03**
Tonne (called “metric ton” in the US) (t)	Kilogram (kg)	**1.0**	**E+03**

To convert from	To	Multiply by	
Ton of refrigeration (12 000 Btu_{IT}/h)	Watt (W)	3.516 853	E+03
Ton of TNT (energy equivalent)	Joule (J)	**4.184**	**E+09**
Ton, register	Cubic meter (m^3)	2.831 685	E+00
Ton, short (2000 lb)	Kilogram (kg)	9.071 847	E+02
Ton, short, per cubic yard	Kilogram per cubic meter (kg/m^3)	1.186 553	E+03
Ton, short, per hour	Kilogram per second (kg/s)	2.519 958	E−01
Torr (Torr)	Pascal (Pa)	1.333 224	E+02
Watt-hour (W · h)	Joule (J)	**3.6**	**E+03**
Yard (yd)	Meter (m)	**9.144**	**E−01**
Year (365 days)	Second (s)	**3.1536**	**E+07**
Year (sidereal)	Second (s)	3.155 815	E+07
Year (tropical)	Second (s)	3.155 693	E+07

Index

B

C

M

N

O

T

About the Author

R. Matthew Brach, PhD, PE, is an engineering consultant with Engineering Systems Inc. His principal areas of professional activities include vehicle impact analysis, vehicle dynamics, automotive crash reconstruction, and automotive safety. He has a PhD in Mechanical Engineering from Michigan State University, an MS in Mechanical Engineering from the University of Illinois at Chicago, and a BS in Electrical Engineering from the University of Notre Dame.

www.ingramcontent.com/pod-product-compliance
Lightning Source LLC
LaVergne TN
LVHW050910210926
848482LV00012B/396
* 9 7 8 1 4 6 8 6 0 5 9 3 8 *